Selected Papers from the 11th Asian Rock Mechanics Symposium (ARMS 11)

Selected Papers from the 11th Asian Rock Mechanics Symposium (ARMS 11)

Editors

Chun Zhu
Zhigang Tao
Xiaojie Yang
Jianping Sun

MDPI • Basel • Beijing • Wuhan • Barcelona • Belgrade • Manchester • Tokyo • Cluj • Tianjin

Editors
Chun Zhu
Hohai University
China

Zhigang Tao
China University of Mining and
Technology (Beijing)
China

Xiaojie Yang
China University of Mining and
Technology (Beijing)
China

Jianping Sun
Nanyang Technological
University,
Singapore

Editorial Office
MDPI
St. Alban-Anlage 66
4052 Basel, Switzerland

This is a reprint of articles from the Special Issue published online in the open access journal *Energies* (ISSN 1996-1073) (available at: https://www.mdpi.com/journal/energies/special_issues/asian_rock_mechanics_symposium).

For citation purposes, cite each article independently as indicated on the article page online and as indicated below:

LastName, A.A.; LastName, B.B.; LastName, C.C. Article Title. *Journal Name* **Year**, *Volume Number*, Page Range.

ISBN 978-3-0365-3268-4 (Hbk)
ISBN 978-3-0365-3269-1 (PDF)

Contents

Editorial

Challenges and Opportunities in Rock Mechanics and Engineering—An Overview

Chun Zhu [1,*], Xiaojie Yang [2], Zhigang Tao [2] and Jianping Sun [3]

[1] School of Earth Sciences and Engineering, Hohai University, Nanjing 210098, China
[2] State Key Laboratory for Geomechanics & Deep Underground Engineering,
China University of Mining and Technology (Beijing), Beijing 100083, China; yxjcumt@163.com (X.Y.);
taozhigang@cumtb.edu.cn (Z.T.)
[3] School of Civil and Environmental Engineering, Nanyang Technological University,
Singapore 639798, Singapore; SUNJ0030@e.ntu.edu.sg
* Correspondence: zhu.chun@hhu.edu.cn

Citation: Zhu, C.; Yang, X.; Tao, Z.; Sun, J. Challenges and Opportunities in Rock Mechanics and Engineering—An Overview. *Energies* **2022**, *15*, 807. https://doi.org/10.3390/en15030807

Received: 21 January 2022
Accepted: 21 January 2022
Published: 22 January 2022

Publisher's Note: MDPI stays neutral with regard to jurisdictional claims in published maps and institutional affiliations.

1. Introduction

The problem of rock mechanics and engineering is an old and new subject encountered by human beings in their struggle with nature for survival and development. To call it ancient means that it has a long history, whereas calling it brand new refers to the continuous emergence of new problems and new situations in engineering practice, which is quite challenging. With the deepening of human engineering activities, the problems of rock engineering are becoming more and more prominent, and the problems encountered are becoming more and more complex. In the practice of solving complex rock engineering, human beings have summarized many topics that are difficult to explain or solve with classical mechanics.

In order to form a unified understanding of the problems and advanced solutions in complex rock engineering, experts at home and abroad have been able to actively deliver the latest research results to this Special Issue. The Special Issue focusses on advances and innovative research on rock mechanics and rock engineering, and provides a showcase of recent developments and advances in rock mechanics and innovative applications in rock engineering. Since the launch of the Special Issue, a total of 15 well-known scholars have submitted their research work to this Special Issue. After strict quality screening, five high-level papers have been published in the *Energies* journal.

2. Special Issue Content

Wang et al. [1] investigated the deformation behavior of crushed mudstones with different particle sizes under incremental loading with an innovative experimental device that simulated boundary conditions of the GERRF method. The influence of particle size of the crushed mudstones to the generation of lateral stress applied on support structures was concurrently observed and analyzed. Research outputs from the tests showed that: (1) the particle size exerted a significant influence on the accumulated axial deformation, period axial deformation, and lateral stress applied on support structure of crushed rocks; (2) under the same axial stress, the larger the particle size, the smaller the accumulated axial deformation of the crushed rock; (3) two types of periodic stress-strain curves were observed for crushed mudstones in the tests; and (4) the lateral pressure generated by large-size samples was smaller than that of small-size samples.

Li et al. [2] prepared six groups of cemented coal gangue-fly ash backfill (CGFB) samples with varying amounts of kaolin instead of cement, and analyzed their mechanical properties using uniaxial compression, acoustic emission, scanning electron microscopy, X-ray diffraction, and Fourier transform infrared spectroscopy. The results show that the uniaxial compressive strength, peak strain, and elastic modulus of CGFB samples decreased with the kaolin content. The average uniaxial compressive strength, elastic modulus, and peak strain of CGFB samples with 10% amount of kaolin are close to that of CGFB samples with no kaolin. The contribution of kaolin hydration to the strength of the CGFB sample is lower than that of cement hydration, and the hydration products, such as ettringite and calcium–silicate–hydrate gel, decrease, thereby reducing strength, which mainly plays a role in filling pores. The contents of kaolin affects the failure characteristics of CGFB samples, which show tensile failure accompanied by local shear failure, and the failure degree increases with the kaolin content.

Liu et al. [3] proposed a nonlinear dynamic simulation method for the rock–fault contact system, and influences of fault dislocations on tunnel stability under seismic action was explored. First, considering the deterioration effect of seismic action on the ultimate bearing load of the contact interface between rock mass and fault, a mathematical model is established reflecting the seismic deterioration laws of the contact interface. Then, based on the traditional point-to-point contact type in a geometric mesh, a point-to-surface contact type is also considered, and an improved dynamic contact force method is established, which considers the large sliding characteristics of the contact interface. Finally, a three-dimensional calculation model for a deep tunnel through a normal fault is built, and the nonlinear seismic damage characteristics of the tunnel under horizontal seismic action are studied. The results indicate that the relative dislocation between the rock mass and the fault is the main factor that results in lining damage and destruction.

Yin et al. [4] performed a dynamic analysis of the slope topography to elaborate on the influences of the directions of seismic waves. Seismic waves were input using an equivalent nodal force method combined with a viscous-spring artificial boundary. The amplification of ground motions in double-faced slope topographies was discussed by varying the angles of incidence. Meanwhile, the components of seismic waves (P waves and SV waves), slope materials, and slope geometries were all investigated with various incident earthquake waves. The results indicated that the pattern of the amplification of SV waves was stronger than that of P waves in the slope topography, especially in the greater incident angels of the incident waves. Soft materials intensely aggravate the acceleration amplification, and more scattered waves are produced under oblique incident earthquake waves. The variations in the acceleration amplification ratios on the slope crest were much more complicated at oblique incident waves, and the ground motions were underestimated by considering only the vertical incident waves.

Ruan et al. [5] performed an analysis of amplitude variation with offset (AVO). Based on the experimental and log data, sensitivity analysis was performed to sort out the rock physics attributes sensitive to microcrack and total porosities. The Biot–Rayleigh poroe-lasticity theory described the complexity of the rock and yielded the seismic properties, such as Poisson's ratio and P-wave impedance, which are used to build rock physics templates calibrated with ultrasonic data at varying effective pressures. The templates were then applied to seismic data of the Xujiahe formation to estimate the total and microcrack porosities, indicating that the results are consistent with actual gas production reports.

3. Closing Remarks

The papers presented in the Special Issue cover important aspects of the latest research progress in rock mechanics and engineering. Even if rock mechanics and engineering is a very wide topic, this small contribution could stimulate the community to develop current research and improve its progress. Therefore, we believe that the presented papers will have practical importance for the future development in the rock mechanics and engineering sector. Finally, together with other co-Guest Editors, Prof. Xiaojie Yang, Prof. Zhigang Tao, and Dr. Jianping Sun, we wish to thank the authors that contributed with their works to this Special Issue.

Funding: This research received no external funding.

Acknowledgments: Each paper was reviewed by at least three reviewers; therefore, the Guest Editors would like to thank them for their work, which helped the authors to improve their manuscripts. In addition, the Guest Editors would like to thank Sherwin Chen for his help and support in the Editorial process of this Special Issue in the Energies journal.

Conflicts of Interest: The authors declare no conflict of interest.

References

1. Wang, Q.; Guo, Z.; Zhu, C.; Yin, S.; Yin, D. The Deformation Characteristics and Lateral Stress of Roadside Crushed Rocks with Different Particles in Non-Pillar Coal Mining. *Energies* **2021**, *14*, 3762. [CrossRef]
2. Li, F.; Yin, D.; Zhu, C.; Wang, F.; Jiang, N.; Zhang, Z. Effects of Kaolin Addition on Mechanical Properties for Cemented Coal Gangue-Fly Ash Backfill under Uniaxial Loading. *Energies* **2021**, *14*, 3693. [CrossRef]
3. Liu, G.; Zhang, Y.; Ren, J.; Xiao, M. Seismic Response Analysis of Tunnel through Fault Considering Dynamic Interaction between Rock Mass and Fault. *Energies* **2021**, *14*, 6700. [CrossRef]
4. Yin, C.; Li, W.-H.; Wang, W. Evaluation of Ground Motion Amplification Effects in Slope Topography Induced by the Arbitrary Directions of Seismic Waves. *Energies* **2021**, *14*, 6744. [CrossRef]
5. Ruan, C.; Ba, J.; Carcione, J.M.; Chen, T.; He, R. Microcrack Porosity Estimation Based on Rock Physics Templates: A Case Study in Sichuan Basin, China. *Energies* **2021**, *14*, 7225. [CrossRef]

Article

The Deformation Characteristics and Lateral Stress of Roadside Crushed Rocks with Different Particles in Non-Pillar Coal Mining

Qiong Wang [1], Zhibiao Guo [2,*], Chun Zhu [3], Songyang Yin [2] and Dawei Yin [4,*]

[1]   Department of Emergency Technology and Management, North China Institute of Science & Technology, Beijing 101601, China; wangqiong@ncist.edu.cn

[2]   School of Mechanics and Civil Engineering, China University of Mining & Technology (Beijing), Beijing 100083, China; BQT1900603026@cumtb.com

[3]   School of Earth Sciences and Engineering, Hohai University, Nanjing 210098, China; zhu.chun@hhu.edu.cn

[4]   State Key Laboratory of Mine Disaster Prevention and Control, Shandong University of Science and Technology, Qingdao 266590, China

*   Correspondence: tsp140101031@student.cumtb.edu.cn (Z.G.); yindawei@sdust.edu.cn (D.Y.); Tel.: +86-139-1028-3906 (Z.G.)

Citation: Wang, Q.; Guo, Z.; Zhu, C.; Yin, S.; Yin, D. The Deformation Characteristics and Lateral Stress of Roadside Crushed Rocks with Different Particles in Non-Pillar Coal Mining. *Energies* **2021**, *14*, 3762. https://doi.org/10.3390/en14133762

Academic Editor: Manoj Khandelwal

Received: 13 May 2021
Accepted: 22 June 2021
Published: 23 June 2021

Publisher's Note: MDPI stays neutral with regard to jurisdictional claims in published maps and institutional affiliations.

Abstract: Gob-side entry retaining formed by roof fracturing (GERRF) is a popular non-pillar mining method. The method uses crushed rocks in gob side to support and control the movements of the gob roof. These crushed rocks will deform under roof pressure and generate desirable lateral stress on support structures of gangue rib. In this study, the deformation behavior of crushed mudstones with different particle sizes under incremental loading was investigated with an innovative experimental device that simulated boundary conditions of the GERRF method. Influence of particle size of the crushed mudstones to the generation of lateral stress applied on support structures were concurrently observed and analyzed. Research outputs from the tests showed that: (1) The particle size exerted a significant influence on the accumulated axial deformation, period axial deformation, and lateral stress applied on support structure of crushed rocks. (2) Under the same axial stress, the larger the particle size, the smaller the accumulated axial deformation of the crushed rock; A skeletal loading-bearing effect was apparent in the rock samples with larger particles (S-2, S-3). The compressive deformation process of samples S-2, S-3 divided into structural adjustment, skeletal load-bearing and crushing cum filling phases. At skeletal loading-bearing phase, the crushed rocks showed better deformation resistance and stability than other phases; (3) Two types of periodic stress-strain curves were observed for crushed mudstones in the tests. The "down-concave" type implied the deformation for the crushed mudstones was primarily a consequence of the compression in the void spaces. While the "upper-convex" type curve was resulted in particle crushing cum filling again; (4) The lateral pressure generated by large-size samples was smaller than that of small-size samples. Additionally, a poor regularity of lateral stress was observed in compression test of large-size sample (S-3). The relationship between the axial stress and lateral stress generated on the support structure was found to be approximately linear relationship under the condition that lateral pressure shows good regularity.

Keywords: non-pillar coal mining; crushed rocks; compression test; deformation characteristics; lateral stress

1. Introduction

Gob-side entry retaining is one of the most commonly used mining method in non-pillar coal mining. In this method, the former entry is artificially retained as the tailgate for the next mining panel by constructing a filling wall made of concrete blocks, pigsties, high-water packing material, and other fill materials, which can greatly improve resource recovery and reduce roadway drivage rate [1,2]. However, in case of complex geological mining engineering, the conventional gob-side entry retaining method encounters inevitable difficulties, due to high stress, high dynamic disturbance and large deformation issues [3–5]. Furthermore, the high cost of filling materials and time-consuming have also severely restricted the wide application of the conventional gob-side entry retaining method.

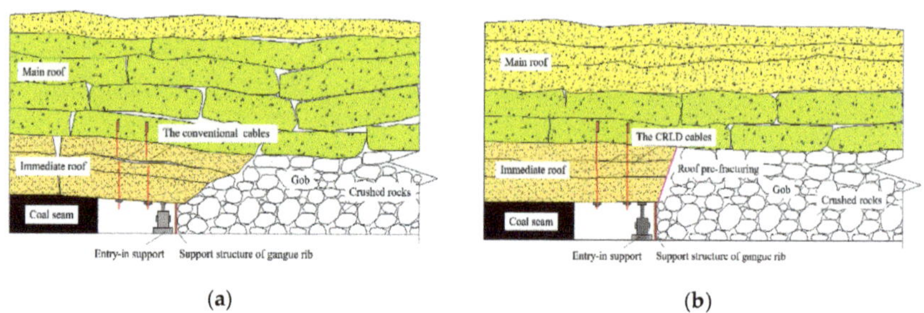

(**a**) (**b**)

Figure 1. Principle of the GERRF method: (**a**) No roof pre-fracturing; (**b**) With roof pre-fracturing.

The crushed rocks are the maintenance body for the GERRF method, its compression and deformation characteristics appear to be very important on the stability of the retained entry. Much research has been focused on the instantaneous compressible deformation [13–15] and creep deformation [16–18] of crushed rocks by uniaxial compression test. However, in these uniaxial compression tests, the crushed rocks were completely constrained on all sides, the boundary conditions were significantly different from the GERRF method [19,20]. Therefore, an innovative experimental device to simulate the boundary condition of crushed rocks in gob area of GERRF was developed. Using the device, the compression tests of crushed mudstones with different particle sizes were carried out. In the tests, the deformation behavior of crushed mudstones with different particle sizes in GERRF method was studied, together with that of the lateral stress giving rise to lateral deformation of support structure was measured, which expects to provide experimental evidence for deformation prediction and supporting design of the GERRF method.

2. An Innovative Experimental Device

In the previous research data obtained from uniaxial compression tests, crushed rocks achieved a greater compactness under axial stress. However, the crushed rocks in the GERRF method are difficult to compact tightly, but rather acquire a new equilibrium state with the surrounding rock. Therefore, an innovative experimental device simulated the geometric structure of the GERRF method was developed, as shown in Figure 2. The experimental device comprises of a loading plate, a cubic frame containing and a base plate. The internal dimensions of the device are 400 mm × 400 mm × 400 mm. Three of the vertical sides of cube are fabricated with Q235 solid steel, while the other side is the simulated surface of gangue rib, which comprises of high-strength wire mesh and scaled down support bars. In accordance with the adopted geometrical similarity ratio of 1:10, the dimensions of the scaled down support bar are 440 mm in height, 10 mm in width and 5 mm in thickness. A total of eight scaled down support bars are arranged on the simulation surface of gangue rib. The arrangement spacing is about 55–56 mm.

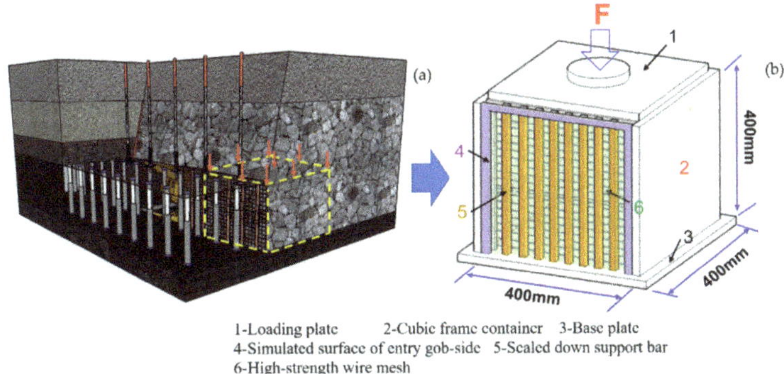

Figure 2. Design principle of the innovative experimental device: (**a**) Schematic diagram of the structure of the GERRF method; (**b**) The structural composition of the experimental device.

The MTS hydraulic servo loading system used in this study is illustrated in Figure 3. The maximum axial load of the triaxial loading system is 2000 kN, and the accuracy of the measurement of axial load is less than 0.01 kN.

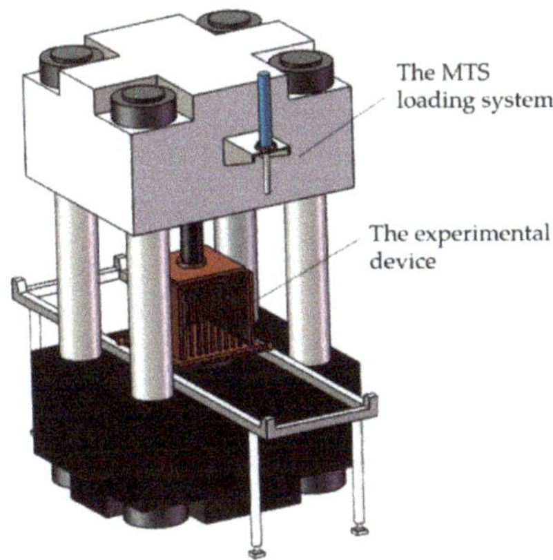

Figure 3. The MTS hydraulic servo loading system.

3. Experimental Materials and Methods

3.1. Crushed Rock Samples

The crushed rock samples used in the tests were the crushed mudstones obtained from the gob roof of 12201 working face in the Haragou Coal Mine (Shenmu, China). The natural density and the uniaxial compressive strength of the crushed rock samples were 2.65 g/cm^3 and 19.8 MPa respectively. In accordance with the need to comply with acceptable size of the particles of crushed rocks in reliable testing, a ratio of 1/5 between the inscribed circle diameter (D) of cubic container [21], the largest crushed rock particles had to be maintained. Therefore, the maximum particle size of samples used in tests was limited to 80 mm. Three types of crushed mudstone samples with different ranges of particle size categories (10~30, 30~60, 60~80 mm) were prepared for the tests, as shown in Figure 4.

(**a**) 10~30 mm   (**b**) 30~60 mm   (**c**) 60~80 mm

Figure 4. Crushed mudstone samples for the tests.

3.2. Lateral Stress Monitoring Plan

To monitor the lateral stress exerted by the crushed rocks on the support structure of gangue rib during the tests, strain gauges were affixed on the scaled down support bars of the simulation surface of gangue rib. In consideration that a certain amount of deformation space was needed to be provided for the axial compression of the crushed mudstones, the strain gauges were so affixed evenly within 230 mm in the vertical direction, as shown in Figure 5.

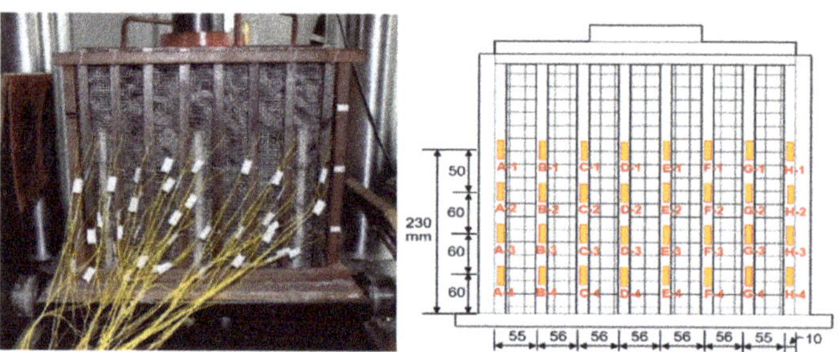

Figure 5. The specific locations and numbering of the strain gauges in tests.

3.3. Loading Scheme

Assuming that the roof pressure acting on the crushed rocks in the gob area remains unchanged during the two caving activities of roof strata, it was considered necessary that the axial pressure acting on the crushed rocks in the gob area to be increased in a stepped manner [22]. Thus, the tests were conducted under incremental loading as shown in Figure 6. In GERRF method, the support structures of gangue rib deformed in response to lateral stress generated by crushed rocks in gob side. In engineering practice, the support structures of gangue rib are not allowed to produce large deformations. So, in the tests, the maximum axial pressure was set at a value corresponding to when the support structure will begin to deform, which was set as 1.5 MPa according to preliminary experiments results. In order to observe the periodic deformation behavior of crushed rocks in greater detail, the loading increment ΔP was designed as 0.15 MPa.

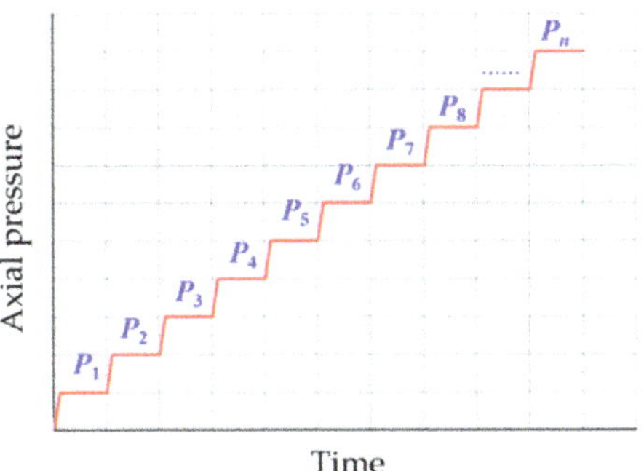

Figure 6. Loading path in the tests.

3.4. Experimental Procedure

The height of the crushed mudstone samples was set to 380 mm before the initiation of the compression tests. The experimental procedures are described as follows:

- Fixed the strain gauges on the inner wall of the support bars.
- Placed the crushed mudstone rocks in cubic container layer by layer, ensuring that the samples were mixed thoroughly. After the height of the samples reaches the target height of 380 mm, the loading plate was assembled and the experimental device was prepared to loaded.
- An initial preload of 20 kN was applied prior to the first compression stage, in order to be rid of any large voids that would still be present in the sample.
- Then the axial loading program was commenced. The axial loading at each stage was completed in 30 s. After the target axial load at each stage was reached, the target stage load was maintained for 10 min before the next loading stage. The loading sequence was terminated when the supporting structure of gangue rib began to deform.

4. Experimental Results

4.1. Axial Deformation

4.1.1. Accumulated Axial Deformation

The stress-strain curves of the crushed mudstone samples for the entire loading process under the incremental loading were recorded in Figure 7. The figure clearly shows that the accumulated axial strain of the crushed mudstone samples increased with increasing axial stress. It can be seen that the accumulated axial strain of S-1, S-2, and S-3 were 14.5%, 10.64%, and 8% respectively when the axial stress $P = 0.6$ MPa (the fourth loading stage); When the axial stress $P = 0.9$ MPa (the sixth loading stage), the accumulated axial strain of S-1, S-2, and S-3 were 17.4%,13.93% and 10.1%. These results indicate that the accumulated axial strain of the samples decreased with increasing particle size under the same axial stress. Additionally, it was noted that the ultimate loading stage for S-1, S-2, S-3 was the 7th, 8th and 9th stage, respectively, which demonstrated that the lateral stress generated on the support structure by the crushed mudstones decreased as the particle size increased under the same axial stress as well. This conclusion will be explained precisely in Section 4.2.

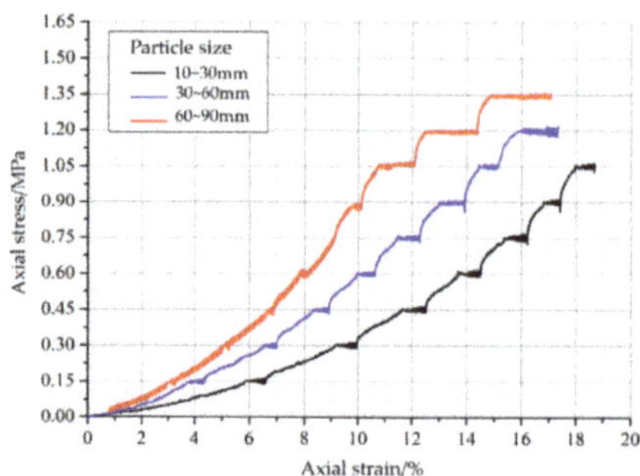

Figure 7. The stress-strain curves of the tested crushed mudstone samples for the entire loading process under the incremental loading.

4.1.2. Periodic Axial Deformation

Figure 8 shows the change rule of axial strain increment corresponding to each loading stage (periodic axial stain) for S-1, S-2 and S-3. And the periodic axial stain values of each loading stage are listed in Table 1. For sample S-1 (Figure 8a), the periodic axial stain decreased with the loading stage increasing; In the case of samples S-2 and S-3, the periodic axial strain showed a decreasing trend in the early loading stages, but rebounded in the later loading stages, as seen in Figure 8b,c. This results indicated that there was a skeletal load-bearing effect in large-sized rushed mudstone samples (S-2, S-3).

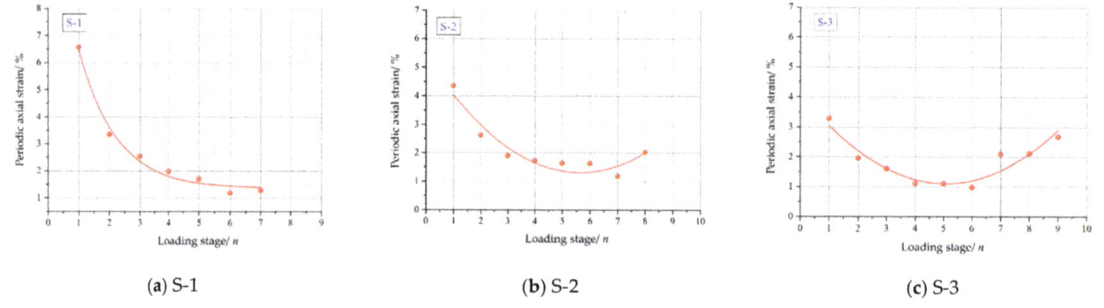

(a) S-1   (b) S-2   (c) S-3

Figure 8. The change rule of axial strain increment corresponding to each loading stage for S-1, S-2 and S-3.

Table 1. The periodic axial strain of crushed mudstone sample at each loading stage.

Particle Size	The Periodic Axial Strain ε (%)								
	1st Stage	2nd Stage	3rd Stage	4th Stage	5th Stage	6th Stage	7th Stage	8th Stage	9th Stage
S-1 (10–30)	6.59	3.37	2.55	1.99	1.71	1.19	1.28	/	/
S-2 (30–60)	4.36	2.64	1.91	1.73	1.65	1.64	1.19	2.04	/
S-3 (60–80)	3.30	1.97	1.61	1.12	1.11	0.99	2.11	2.14	2.69

For crushed mudstone sample S-2 and S-3, the entire deformation process can be divided into structural adjustment, skeleton load-bearing and crushing cum filling phases, as presented in Figure 9. In the skeleton load-bearing phase, the periodic axial strain was relatively smaller, which indicated that the crushed rocks with large-sized particles have higher deformation resistance and stability in the skeleton load-bearing phase than other phases.

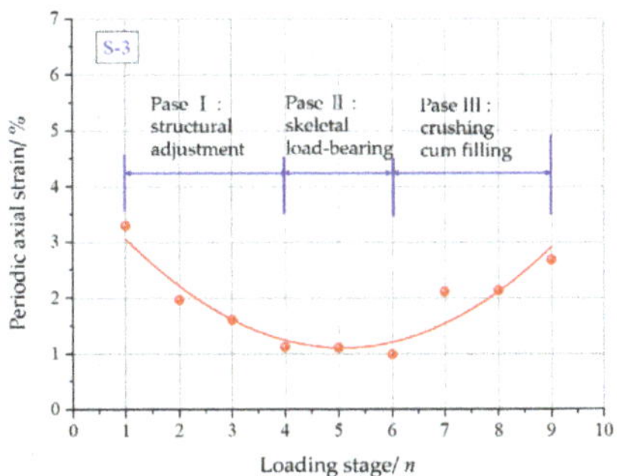

Figure 9. The entire deformation process of the crushed rock sample with skeleton load-bearing effect.

The periodic axial stress-strain curves of the crushed mudstone samples at each loading stage were presented in Figures 10–12. From these figures, we known that the periodic axial strain of the crushed mudstones was composed of instantaneous deformation and creep deformation. Observations of the instantaneous axial strain and axial creep strain of crushed mudstones at each loading stage were listed in Tables 2 and 3 respectively. Table 2 showed that the instantaneous axial strain of all samples decreased with increasing loading stage. From observations in Table 3, we know that the axial creep strain of sample S-1 at each loading stage was approximately similar, and most of them were between 0.7–0.8. But for sample S-2 and S-3, the axial creep strain was a minimum at the skeletal load-bearing stage. As it entered the crushing cum filling stage, the periodic creep deformation increased sharply.

Table 2. The periodic axial instantaneous strain of crushed mudstone samples at each loading stage.

Particle Size	The Periodic Axial Instantaneous Strain ε_1 (%)								
	1st Stage	2nd Stage	3rd Stage	4th Stage	5th Stage	6th Stage	7th Stage	8th Stage	9th Stage
S-1 (10–30)	5.80	2.63	1.75	1.20	0.88	0.6	0.56	/	/
S-2 (30–60)	3.80	2.22	1.33	1.05	0.81	0.7	0.58	0.45	/
S-3 (60–80)	3.00	1.70	1.34	0.90	1.10	0.55	0.63	0.40	0.25

Table 3. The periodic creep strain of crushed mudstone samples at each loading stage.

Particle Size	The Periodic Creep Strain ε_2 (%)								
	1st Stage	2nd Stage	3rd Stage	4th Stage	5th Stage	6th Stage	7th Stage	8th Stage	9th Stage
S-1 (10–30)	0.79	0.74	0.80	0.79	0.83	0.59	0.72	/	/
S-2 (30–60)	0.56	0.42	0.58	0.68	0.84	0.94	0.61	1.59	/
S-3 (60–80)	0.30	0.27	0.27	0.22	0.01	0.44	1.48	1.74	2.44

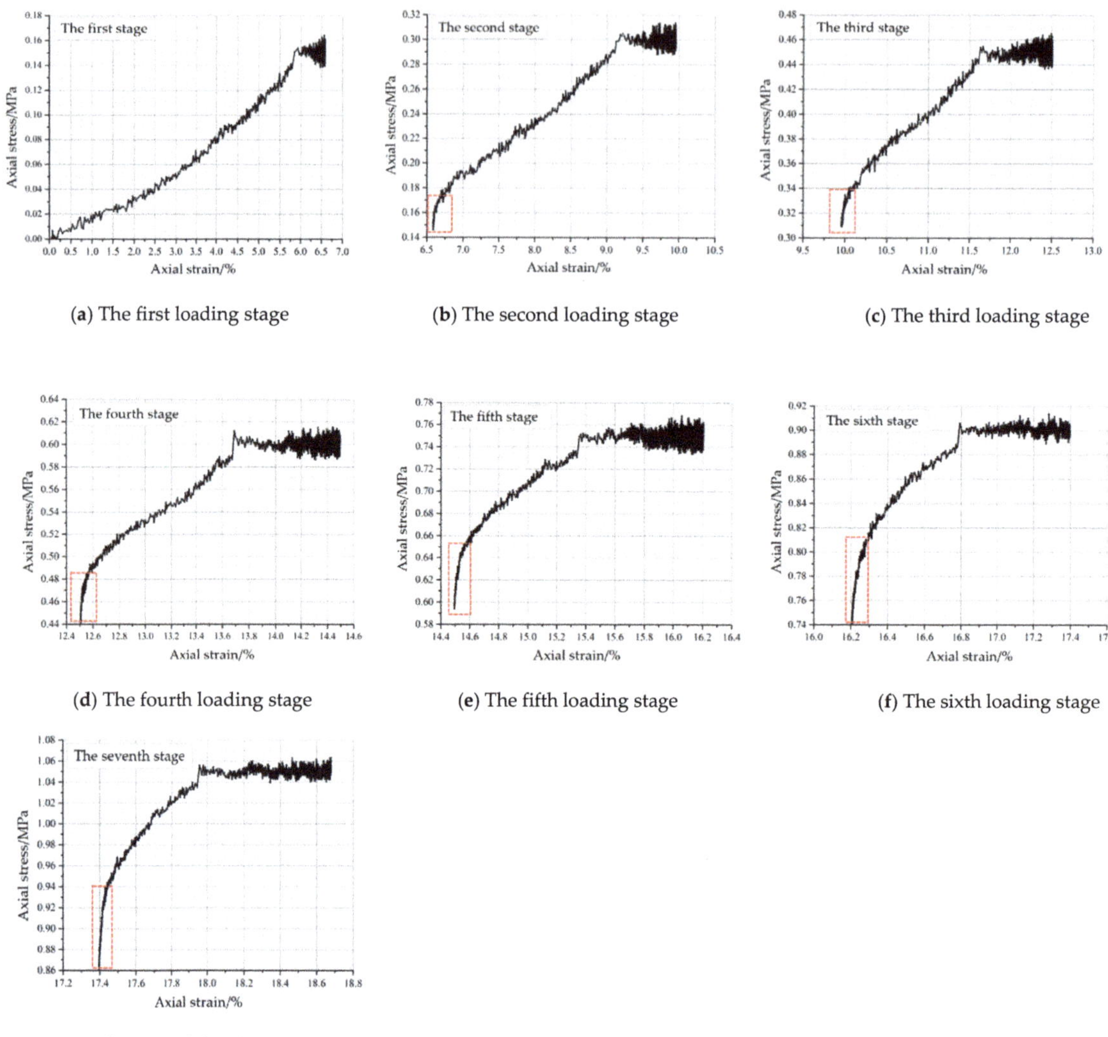

(**a**) The first loading stage

(**b**) The second loading stage

(**c**) The third loading stage

(**d**) The fourth loading stage

(**e**) The fifth loading stage

(**f**) The sixth loading stage

(**g**) The seventh loading stage

Figure 10. The periodic stress-strain curves of sample S-1 at each loading stage.

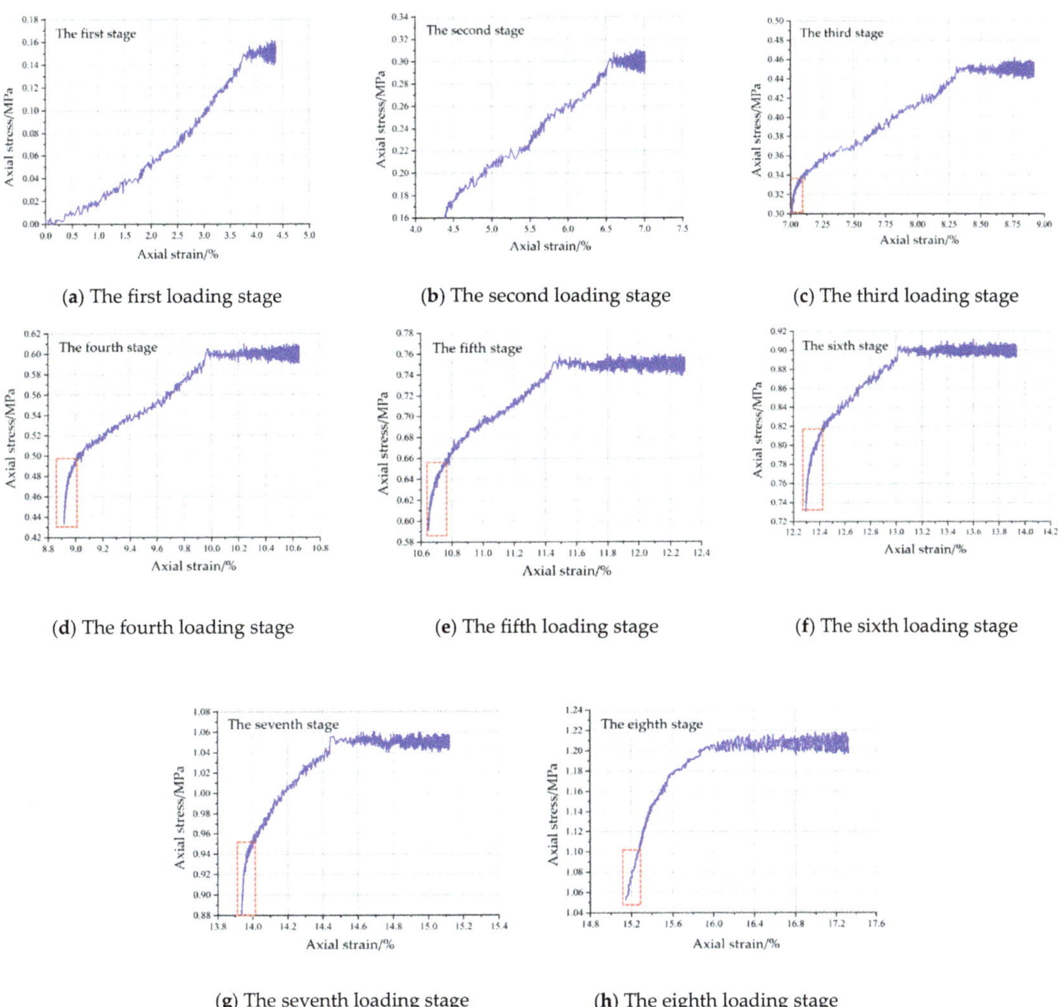

(**a**) The first loading stage

(**b**) The second loading stage

(**c**) The third loading stage

(**d**) The fourth loading stage

(**e**) The fifth loading stage

(**f**) The sixth loading stage

(**g**) The seventh loading stage

(**h**) The eighth loading stage

Figure 11. The periodic stress-strain curves of sample S-2 at each loading stage.

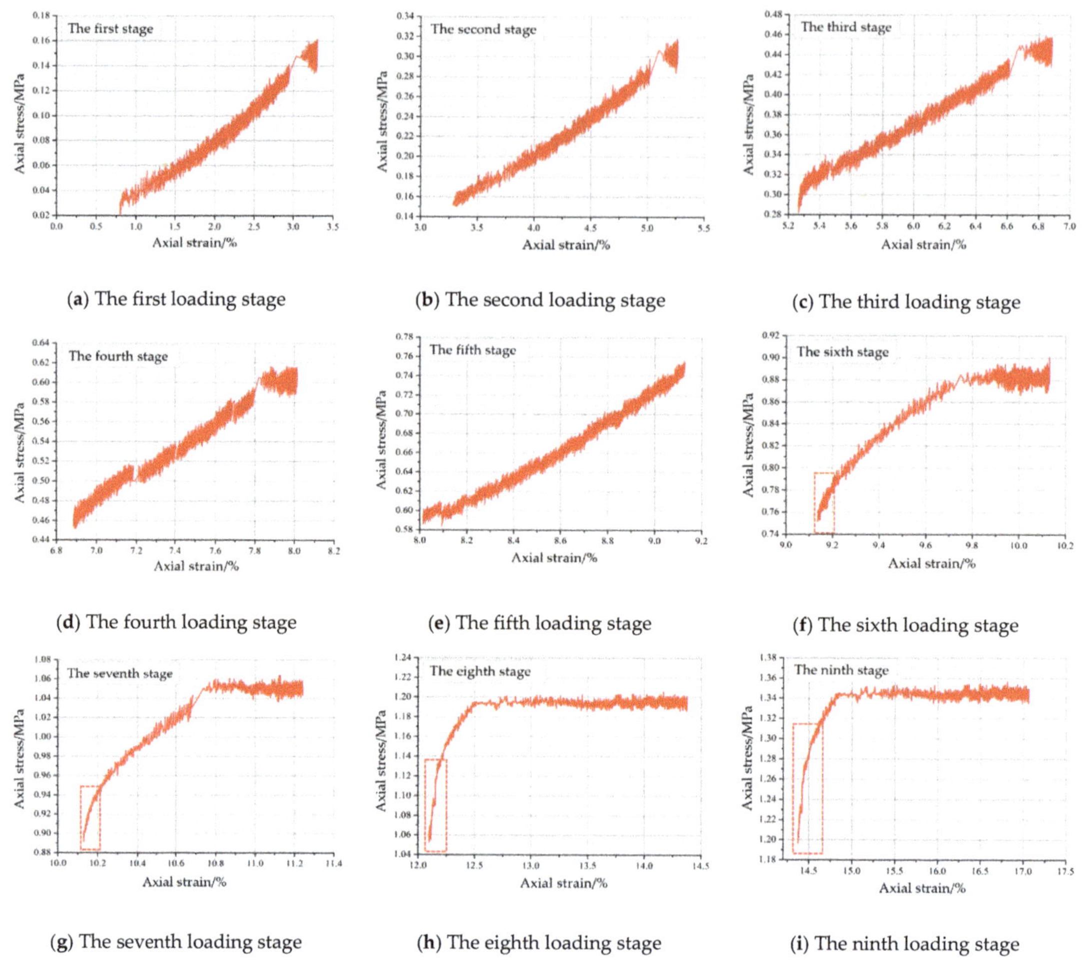

Figure 12. The periodic stress-strain curves of sample S-3 at each loading stage.

In addition, two types of periodic stress-strain curves were observed from the test results, which were "down-concave" type and "upper-convex" type, respectively. The "down-concave" type stress-strain curve appeared at the previous loading stages, which presented a large growth rate of axial strain at the earlier phases of the loading stage, and a smaller growth rate at the later phases of the loading stage. Conversely, for the "upper-concave type" stress-strain curve, the growth rate of axial strain at earlier phases of the loading stage was small (shown within a red colored frame in Figures 10–12). But after the axial stress reached a certain value, the axial strain increased continuously.

4.2. Lateral Stress

Figure 13 shows the lateral stress of the D column monitoring points on the simulated support structure in the tests. From the results, it can be seen that the lateral stress increased as the axial stress increased. With the increase of particle size, the lateral stress generated on the support structure decreases under the same axial stress. In addition, the test results showed that the lateral stress generated by the sample S-1 and S-2 consistently showed a better regularity than sample S-3.

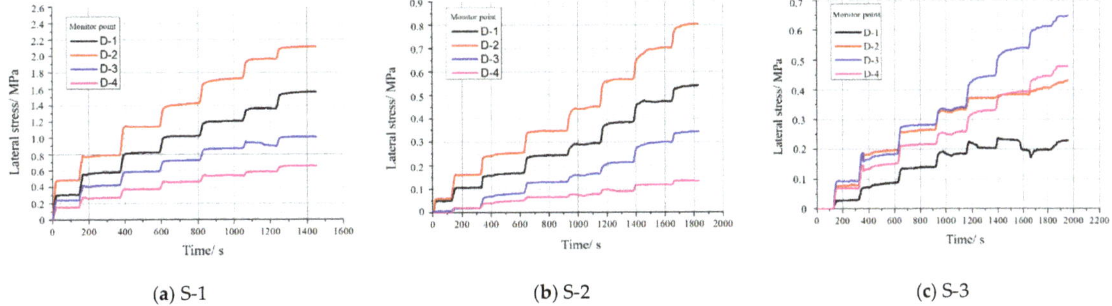

(a) S-1   (b) S-2   (c) S-3

Figure 13. The lateral stress of the D column monitoring points.

Using a curve fitting analysis, a linear relationship between the lateral stress generated by crushed mudstone rocks and the axial stress can be represented by the following equation:

$$\sigma_x = a\sigma_y + b \tag{1}$$

where σ_x is the lateral stress from the crushed mudstones that was generated on the support structure; σ_y is the axial stress applied on crushed mudstones; a and b are regression coefficients. The regression coefficients established for all the tests are itemized in Table 4. The most of correlation coefficients are greater than 0.95.

Table 4. Regression coefficients of the linear relationship between axial stress and lateral stress generated by crushed mudstones.

Particle Size	Monitor Point	Regression Coefficient		Correlation Coefficients
		a	*b*	
S-1 (10–30)	D-1	1.41	106.25	0.977
	D-2	2.00	160.00	0.978
	D-3	0.94	99.16	0.963
	D-4	0.64	77.92	0.954
S-2 (30–60)	D-1	0.53	−4.58	0.993
	D-2	0.77	0	0.998
	D-3	0.34	−22.92	0.978
	D-4	0.13	2.92	0.989
S-3 (60–80)	D-1	0.88	64.17	0.884
	D-2	1.58	186.67	0.910
	D-3	2.35	18.33	0.997
	D-4	1.69	45	0.994

5. Discussion

The reasons for deformation of the crushed rocks include void space compression, particle crushing and particle splitting. The different types of periodic stress-strain curves represented different deformation mechanisms for crushed mudstones. As shown in Figure 14a, for "down-concave type" stress-strain curve, the rapid increase of axial strain in the early loading stage was caused by the compression of a large amount of void space. As the void space was primarily compressed and the rate of axial strain increase slowed down, thus making the stress-strain curve of crushed rocks to present a down-concave shape. Figure 14b shows the deformation mechanism for "upper-convex type" stress-strain curve. The crushed rocks with this type stress-strain curve usually indicate that the sample has a certain initial bearing strength. When the axial stress is greater than the bearing strength of the crushed rock sample, a large number of particles crush and fill the void spaces. This provide a good explanation for the rapidly growth of axial strain of crushed rocks in the later loading stage.

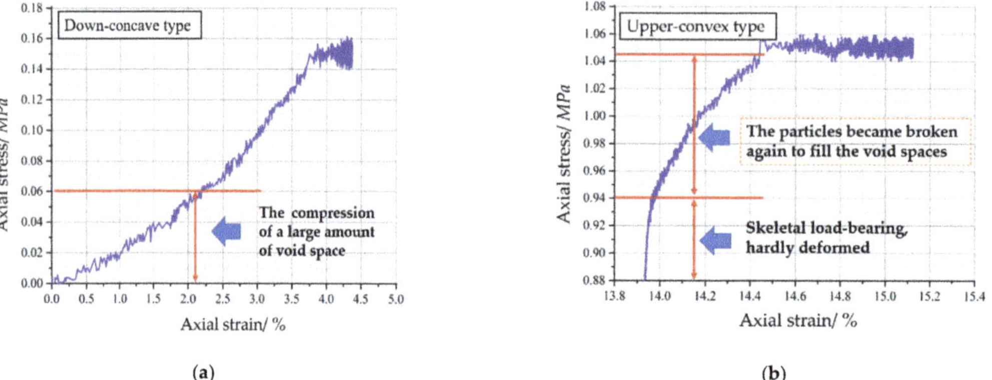

Figure 14. The deformation mechanism of two types of stress-strain curves: (**a**) down-concave type; (**b**) upper-convex type.

From above experimental results, a poor regularity of lateral stress generated by sample S-3 was observed. The phenomenon was illustrated in Figure 15.

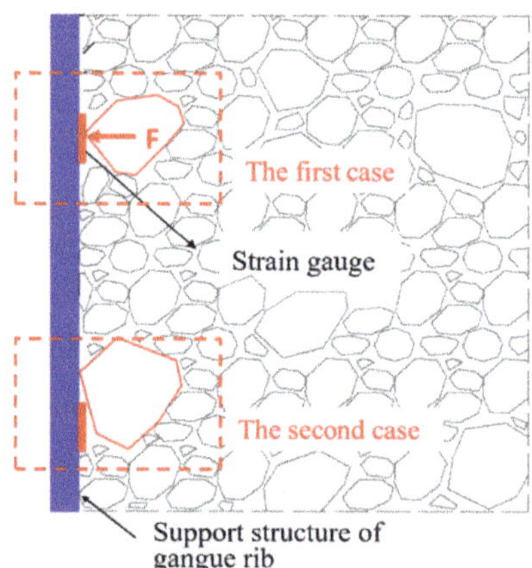

Figure 15. The schematic diagram for a poor regularity of lateral stress generated by larger-sized crushed rocks.

It can be seen that the contact between fine-sized crushed rocks and support structure is more sufficient than the larger-sized crushed rocks. Stress concentration (the first case of Figure 15) and zero-contact (the second case of Figure 15) are more likely to occur in larger-sized crushed rocks.

6. Conclusions

The conclusions drawn from this research are as follows:

(1) An innovative experimental device was developed to simulate the boundary conditions of the GERRF method. Using the device, the compressing tests were conducted

to study the deformation behaviors of crushed rocks with different particle sizes in gob side of GERRF method.

(2) In tests, the accumulated axial deformation of the crushed rocks increased with increasing axial stress. As the particle size increased, the accumulated axial deformation decreased under the same axial stress. In addition, the skeletal loading-bearing effect was found in the samples with larger sized particles. And the entire deformation process of those samples can be divided into structural adjustment, skeleton load-bearing and crushing cum filling phases.

(3) The periodic deformation of the crushed mudstones includes instantaneous compressive deformation and creep deformation. Regardless of the particle size of the crushed rocks, the instantaneous compressive deformation decreased with the increase of loading stage. However, the different change laws of creep deformation were observed in the samples with different particle size. For the sample S-1, the creep deformations at each loading stage were roughly the same. But for samples S-2 and S-3, the creep deformation was minimum at the the skeletal load-bearing phase, but increased when it entered the crushing cum filling phase.

(4) There were two types of periodic stress-strain curves for crushed rocks. The "down-concave" stress-strain curve indicated that the deformation of crushed rocks was mainly caused by the compression of void spaces. While the "upper-convex" curve is the result of particles crushing and particles filling again.

(5) With the increase of particle size, the lateral stress generated on the support structure decreases under the same axial stress. Additionally, a poor regularity of lateral stress generated by crushed rocks with larger- sized particles was observed in tests. Under the condition that lateral pressure shows good regularity, a linear relationship between the axial stress and lateral stress generated by crushed rocks was established to be of the form: $\sigma_x = a\sigma_y + b$.

Author Contributions: All the authors contributed to this paper. Q.W. prepared and edited the manuscript. Z.G. provided theoretical and methodological guidance in the research process. S.Y., C.Z. and D.Y. partially participated in literature search and data processing. All authors have read and agreed to the published version of the manuscript.

Funding: This research was supported by the Science and Technology Project of Langfang City, China (grant number 2020013039), the Special Fund of Basic Research and Operating of North China Institute of Science & Technology (grant number 3142020001), and the State Scholarship Fund of China, the National Natural Science Foundation of China (grant number 51479195).

Institutional Review Board Statement: Not applicable.

Informed Consent Statement: Not applicable.

Data Availability Statement: Not applicable.

Acknowledgments: We are grateful to China University of Mining and Technology for providing us with the experimental platform and all the reviewers for their specific comments and suggestions.

Conflicts of Interest: The authors declare no conflict of interest.

References

1. Zhang, N.; Yuan, L.; Han, C.L.; Xue, J.H.; Kan, J.G. Stability and deformation of surrounding rock in pillarless gob-side entry retaining. *Saf. Sci.* **2012**, *50*, 593–597. [CrossRef]
2. Yang, H.Y.; Cao, S.G.; Wang, S.Q.; Fan, Y.C.; Wang, S.; Chen, X.Z. Adaptation assessment of gob-side entry retaining based on geological factors. *Eng. Geol.* **2016**, *209*, 143–151. [CrossRef]
3. Guo, Z.B.; Wang, Q.; Li, Z.H.; He, M.C. Surrounding Rock Control of An Innovative Gob-side Entry Retaining with Energy-absorbing Supporting in Deep Mining. *Low Carbon Technol.* **2019**, *14*, 23–35. [CrossRef]
4. Wang, Q.; Jiang, Z.H.; Jiang, B.; Gao, H.K.; Huang, Y.B.; Zhang, P. Research on an automatic roadway formation method in deep mining areas by roof cutting with high-strength bolt-grouting. *Int. J. Rock Mech. Min. Sci.* **2020**, *128*, 104264. [CrossRef]
5. Yin, Q.; Wu, J.Y.; Zhu, C.; He, M.C.; Meng, Q.X. Shear mechanical responses of sandstone exposed to high temperature under constant normal stiffness boundary conditions. *Geomech. Geophys. Geo Energy Geo-Resour.* **2021**, *7*, 35. [CrossRef]

6. He, M.C.; Zhu, G.L.; Guo, Z.B. Longwall mining "cutting cantilever beam theory" and 110 mining method in China-The third mining science innovation. *J. Rock Mech. Geotech. Eng.* **2015**, *7*, 483–492. [CrossRef]

7. Wang, Q.; He, M.C.; Li, S.C.; Jiang, Z.H.; Wang, Y.; Qin, Q.; Jiang, B. Comparative study of model tests on automatically formed roadway and gob-side entry driving in deep coal mines. *Int. J. Min. Sci. Technol.* **2021**. [CrossRef]

8. He, M.C.; Gao, Y.B.; Yang, J.; Gong, W.L. An Innovative Approach for Gob-Side Entry Retaining in Thick Coal Seam Longwall Mining. *Energies* **2017**, *10*, 1785. [CrossRef]

9. Yang, X.J.; Wang, E.Y.; Ma, X.G.; Zhang, G.F.; Huang, R.F.; Lou, H.P. A Case Study on Optimization and Control Techniques for Entry Stability in Non-Pillar Longwall Mining. *Energies* **2019**, *12*, 391. [CrossRef]

10. He, M.C.; Gong, W.L.; Wang, J.; Qi, P.; Tao, Z.G.; Du, S.; Peng, Y.Y. Development of a novel energy–absorbing bolt with extraordinarily large elongation and constant resistance. *Int. J. Rock Mech. Min. Sci.* **2014**, *67*, 29–42. [CrossRef]

11. Zhu, C.; He, M.C.; Karakus, M.; Zhang, X.H.; Tao, Z.G. Numerical simulations of the failure process of anaclinal slope physical model and control mechanism of negative Poisson's ratio cable. *Bull. Eng. Geol. Environ.* **2021**, *80*, 3365–3380. [CrossRef]

12. Tao, Z.G.; Zhu, C.; He, M.C.; Karakus, M. A physical modeling-based study on the control mechanisms of Negative Poisson's ratio anchor cable on the stratified toppling deformation of anti-inclined slopes. *Int. J. Rock Mech. Min. Sci.* **2021**, *138*, 104632. [CrossRef]

13. Zhou, N.; Han, X.L.; Zhang, J.X.; Li, M. Compressive deformation and energy dissipation of crushed coal gangue. *Powder Technol.* **2016**, *297*, 220–228. [CrossRef]

14. Ma, D.; Bai, H.B.; Miao, X.X.; Pu, H.; Jiang, B.Y.; Chen, Z.Q. Compaction and seepage properties of crushed limestone particle mixture: An experimental investigation for Ordovician karst collapse pillar groundwater inrush. *Environ. Earth Sci.* **2016**, *75*, 1–12. [CrossRef]

15. Kong, H.L.; Chen, Z.Q.; Wang, L.Z.; Shen, H.D. Experimental study on permeability of crushed gangues during compaction. *Int. J. Miner. Process.* **2013**, *124*, 95–101. [CrossRef]

16. Ma, Z.G.; Gu, R.X.; Huang, Z.M.; Peng, G. Experimental study on creep behavior of saturated disaggregated sandstone. *Int. J. Rock Mech. Min. Sci.* **2014**, *66*, 76–83. [CrossRef]

17. Su, C.L.; Li, G.C.; Gomah, M.E.; Xu, J.H.; Sum, Y.T. Creep characteristics of coal and rock investigated by nanoindentation. *Int. J. Rock Mech. Min. Sci.* **2020**, *30*, 769–776.

18. Guo, Z.B.; Wang, Q.; Yin, S.Y. The creep compaction behavior of crushed mudstones under the step loading in underground mining. *Int. J. Coal Sci. Technol.* **2019**, *6*, 408–418. [CrossRef]

19. Yu, B.Y.; Chen, Z.Q.; Wu, J.Y. Experimental study on compaction and fractal characteristics of saturated broken rocks with different initial gradations. *J. Min. Saf. Eng.* **2016**, *33*, 342–347.

20. Ma, Z.G.; Guo, G.L.; Chen, R.H.; Mao, X.B. An experimental study on the compaction of water-saturated over-broken rock. *Chin. J. Rock Mech. Eng.* **2004**, *25*, 1139–1144.

21. Zhang, J.X.; Li, M.; Liu, Z. Fractal characteristics of crushed particles of coal gangue under compaction. *Powder Technol.* **2017**, *305*, 12–18. [CrossRef]

22. Zhang, J.W.; Wang, H.L.; Chen, S.J.; Li, Y.L. Bearing deformation characteristics of large-size broken rock. *J. China Coal Soc.* **2018**, *43*, 1000–1007.

Article

Effects of Kaolin Addition on Mechanical Properties for Cemented Coal Gangue-Fly Ash Backfill under Uniaxial Loading

Faxin Li [1], Dawei Yin [1,*], Chun Zhu [2], Feng Wang [1], Ning Jiang [1] and Zhen Zhang [1]

[1] State Key Laboratory of Mine Disaster Prevention and Control, Shandong University of Science and Technology, Qianwangang Road 579, Qingdao 266590, China; lfxsdkjdx@163.com (F.L.); wangfeng@sdust.edu.cn (F.W.); jiangning@sdust.edu.cn (N.J.); 13573247170@163.com (Z.Z.)
[2] School of Earth Sciences and Engineering, Hohai University, Xikang Road 1, Nanjing 210098, China; zhu.chun@hhu.edu.cn
* Correspondence: yindawei@sdust.edu.cn; Tel.: +86-18765927224

Citation: Li, F.; Yin, D.; Zhu, C.; Wang, F.; Jiang, N.; Zhang, Z. Effects of Kaolin Addition on Mechanical Properties for Cemented Coal Gangue-Fly Ash Backfill under Uniaxial Loading. *Energies* **2021**, *14*, 3693. https://doi.org/10.3390/en14123693

Academic Editor: Sarma V. Pisupati

Received: 14 March 2021
Accepted: 17 June 2021
Published: 21 June 2021

Publisher's Note: MDPI stays neutral with regard to jurisdictional claims in published maps and institutional affiliations.

Abstract: In this investigation, six groups of cemented coal gangue-fly ash backfill (CGFB) samples with varying amounts of kaolin (0, 10, 20, 30, 40, and 50%) instead of cement are prepared, and their mechanical properties are analyzed using uniaxial compression, acoustic emission, scanning electron microscopy, X-ray diffraction, and Fourier transform infrared spectroscopy. The uniaxial compressive strength, peak strain, and elastic modulus of CGFB samples decreased with the kaolin content. The average uniaxial compressive strength, elastic modulus, and peak strain of CGFB samples with 10% amount of kaolin are close to that of CGFB samples with no kaolin. The contribution of kaolin hydration to the strength of CGFB sample is lower than that of cement hydration, and the hydration products such as ettringite and calcium-silicate-hydrate gel decrease, thereby reducing strength, which mainly plays a role in filling pores. The contents of kaolin affect the failure characteristics of CGFB samples, which show tensile failure accompanied by local shear failure, and the failure degree increases with the kaolin content. The porosity of the fracture surface shows a decreasing trend as a whole. When the amount of kaolin instead of cement is 10%, the mechanical properties of CGFB samples are slightly different from those of CGFB samples without kaolin, and CGFB can meet the demand of filling strength. The research results provide a theoretical basis for the application of kaolin admixture in fill mining.

Keywords: kaolin; cemented coal gangue-fly ash backfill sample; mechanical properties; macroscopic failure; microstructure

1. Introduction

Mine-filling technology processes waste such as coal gangue and fly ash into slurry, which not only reduces the accumulation of solid waste and pollution in coal mines, but also enables effective control of the deformation of overlying strata and surface subsidence in mining areas [1]. However, the application of this technology in coal mines is limited owing to insufficient availability of backfill materials and high costs [2,3]. Therefore, cheap, efficient, and widely available backfill materials are required for mine-filling in the future.

China is rich in kaolin resources and has suitable conditions for development and utilization of kaolin. Kaolin ($Al_2O_3 \cdot 2SiO_2 \cdot 2H_2O$) is a type of fine and soft clay mineral having a "single net layer" structure [4–7]. At present, scholars in China and abroad are studying the influences of kaolin admixtures on mechanical properties and the microstructure of concrete materials. Due to the low activity of kaolin, Chun et al. [8] stimulated the pozzolanic activity of kaolin using high-temperature calcination and found that adding an appropriate amount of kaolin into concrete can greatly improve its compressive strength. Lingyan et al. [9] found that calcined activated kaolin has the greatest effect on the early

19

strength of concrete admixtures, which increases with the addition of kaolin. When the amount of kaolin is 10%, the strength of the concrete admixture is the highest [10–13]. Meng et al. [14] found that the early strength of cement can be improved by mixing calcined activated kaolin and slag. However, the kaolin production via calcination is a high-temperature process that is cumbersome, consumes high energy, and is expensive. Hao [15] and Yuanyuan et al. [16] found that kaolin can react with $Ca(OH)_2$ slowly at room temperature to form hydration products of cementitious ability, and the hydration reaction with cement can create an alkaline environment and enhance this process. Wei et al. [17] found that kaolin and other mineral admixtures can improve the early strength of cement mortar in an alkaline environment. Annan et al. [18] confirmed the plate morphology of kaolin. The particle size of kaolin is mainly in the range of 0–5 μm, which can be filled into smaller pores of cement paste and be dispersed more evenly, which is conducive to the complete occurrence of chemical reactions. Mengna et al. [19] found that the reaction of kaolin and $Ca(OH)_2$ can produce flocculent substances and platelike crystals. The addition of slag and fly ash is conducive to the diffusion of crystals and destroys the structure of $Ca(OH)_2$, thus reducing the porosity of the cement mortar and improving the density of the slurry and filler–matrix interface. Das et al. [20] found that kaolin contains more SiO_2 and Al_2O_3 under alkaline conditions, leading to a higher pozzolanic activity. Its internal structure contains more chemical bonds, which can weaken the secondary hydration of $Ca(OH)_2$ in cement mortar.

The abovementioned research results are of great significance for understanding the effect of kaolin addition on the properties of cement-based materials such as concrete and mortar. The cost of backfilling can be reduced if cement can be replaced with kaolin in cemented coal gangue-fly ash backfill (CGFB) samples. Moreover, it can provide a new path for resource utilization of kaolin. Therefore, based on the test methods of the loading system, acoustic emission (AE), digital video camera (DVC), scanning electron microscope (SEM), X-ray diffraction (XRD), and Fourier-transform infrared spectroscopy (FTIR), we studied the effects of kaolin partially replacing cement on the mechanical properties of CGFB samples, and the feasibility of using kaolin as paste admixture was discussed. The results can provide a theoretical basis for application of kaolin admixture in fill mining.

2. Materials and Methods

2.1. Raw Materials

As shown in Figure 1, the CGFB samples used in this test are composed of cement, fly ash, gangue, kaolin, and water. Their main components and contents are listed in Table 1.

Table 1. Main components and contents.

Raw Material	Chemical Composition and Content/%											
	SiO_2	Al_2O_3	CaO	Na_2O	SO_3	K_2O	MgO	TiO_2	MnO	Fe_2O_3	ZnO	BaO
Cement	15.722	12.340	51.643	2.177	3.025	1.203	8.706	1.031	0.190	3.891	0.072	—
Fly ash	37.855	39.724	6.214	1.456	1.221	2.118	3.914	2.479	0.055	4.856	0.107	—
Gangue	39.987	30.177	11.727	1.758	2.886	2.643	3.088	2.531	—	4.928	—	0.276
Kaolin	42.153	47.704	—	—	—	5.063	1.276	2.079	—	1.725	—	—

Here, 32.5 grade ordinary Portland cement (OPC) produced by Shandong Rizhao No. 3 cement plant was used. Grade II fly ash from the Huangdao Power Plant in Qingdao City, Shandong Province, which has a light grey appearance, was used. The gangue (particle size < 25 mm) was obtained from the solid waste produced in roadway excavation, coal mining, and separation in the Shandong Daizhuang coal mine. The kaolin was taken from the water-washed kaolin produced by a material factory in Guangdong Province. The 80-μm-square pore sieve residue is less than 8%, and the 45-μm-square pore sieve residue is less than 25%, which meets the technical requirements of general Portland cement (GB/T 175-2007) [21].

Figure 1. Raw materials. (**a**) Cement; (**b**) Fly ash; (**c**) Gangue; (**d**) Kaolin.

The microscopic morphology of kaolin was analyzed by a scanning electron microscopy (SEM) system. The powder sample was dipped on the conductive adhesive using a wooden stick, and the kaolin surface was plated with gold. The morphology of the sample was observed at 10,000× magnification, as shown in Figure 2a. The layered structure of kaolin particles is evident in the figure, and each block of kaolin particles is composed of many lamellar structures closely superimposed together, with clear edges and corners. The morphology of kaolin particles is irregular and the particle size is different, which is conducive to backfill the internal pores of CGFB samples, increasing its internal occlusal degree and improving its internal structure.

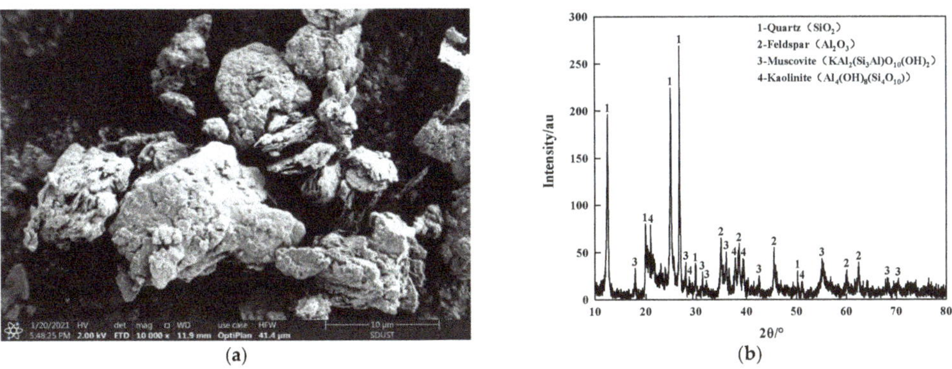

Figure 2. Microstructure analysis of kaolin. (**a**) SEM of kaolin; (**b**) XRD pattern of kaolin.

The crystal phase of the kaolin admixture was analyzed using XRD, as shown in Figure 2b [22]. The XRD patterns show multiple dispersion peaks of quartz, feldspar, muscovite, and kaolinite, among which quartz and feldspar are mainly composed of SiO_2 and Al_2O_3, respectively. The kaolin admixture is rich in active SiO_2 and Al_2O_3, which can replace a part of the cement for pozzolanic reaction and improve the internal bonding of CGFB samples.

2.2. Sample Preparation

In this test, CGFB samples were prepared using cement, fly ash, and gangue in the ratio of 1:4:6; the solid mass fraction was 78%; no additions were added, and the amounts of kaolin replacing cement were 0, 10, 20, 30, 40, and 50%. During the preparation of CGFB samples, an NJ-160 agitator was used for stirring for approximately 8 min. After the slurry was evenly mixed, it was poured into a Φ50 mm $\times$ 100 mm mould. The bubbles in the samples were removed by manual vibration and tamping. CGFB samples were removed from moulds after 24 h and cured for 28 days in a curing box at a temperature of 25 °C and relative humidity of 80%. Before the test, the two ends of the samples were smoothed with a grinding machine: the flatness tolerance of the end face was less than 0.05 mm, and unevenness was less than 0.002 mm [23]. Evident cracks on the surface of the CGFB samples were removed. A total of 18 samples, as shown in Figure 3, were prepared for this test. Based on the amounts of cement replaced with kaolin, they were divided into six groups: A, B, C, D, E, and F, corresponding to cement replacement of 0, 10, 20, 30, 40, and 50% chemically pure (CP), respectively, and each group contained three samples.

Figure 3. CGFB samples.

2.3. Test Method

The experimental test setup, including the loading, AE, DVC, SEM, XRD, and FTIR systems, is shown in Figure 4. During each test, loading, AE, and DVC systems were synchronized to have the same timestamps to facilitate analysis of the experimental results.

2.3.1. Uniaxial Compression Tests

A Shimadzu AG-X250 electronic universal testing machine was used to conduct the uniaxial compression tests on the CGFB samples. This machine can perform uniaxial compression, tensile, and other mechanical tests using a maximum load of 250 kN [24–27]. When performing uniaxial compression tests, a preload pressure of 0.1 kN was first applied to the test sample, so that the indenter was in close contact with the test piece, and displacement loading control at a loading rate of 0.0005 mm/s was conducted.

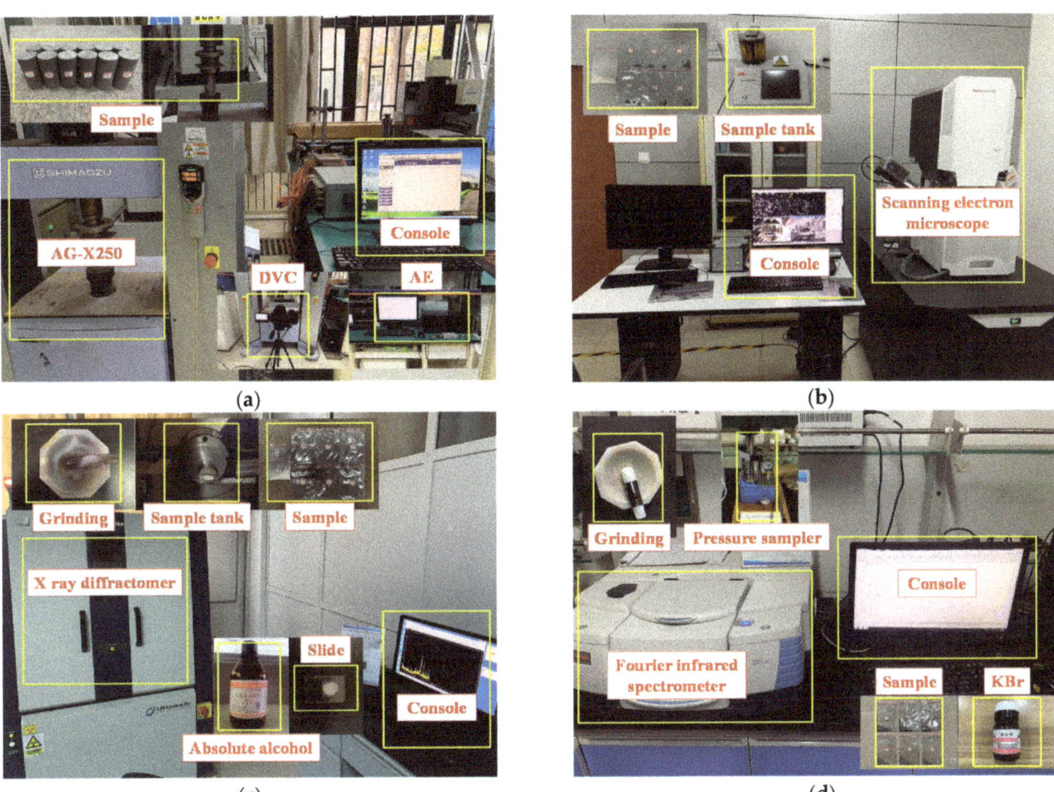

Figure 4. Experimental testing system. (**a**) Uniaxial compression tests; (**b**) SEM experiment; (**c**) XRD experiment; (**d**) FTIR experiment.

2.3.2. Acoustic Emission Experiment

The failure under uniaxial compression was monitored in real-time using the MIS-TRAS series PCI-2 AE system. An R3α type of AE sensor, a main amplifier of 40 dB, threshold of 45 dB, floating threshold of 6 dB, probe harmonic frequency of 100–600 kHz, and sampling frequency of 10^6 times/s were used [28]. Petroleum jelly (Vaseline) was applied between the sensor and samples for coupling them and reducing the acoustic impedance difference and reflection loss of energy at the interface. This ensured that the sensor received the AE signal with minimal loss. The sensor was fixed with adhesive tape, and the pencil lead fracture method proposed by American Society for Testing and Materials (ASTM) was used to calibrate the AE system for ensuring that the signal amplitude of each sensor was above 90 dB [29]. During the test, a video camera (Sony DVC) was used to record the failure under uniaxial loading.

2.3.3. Scanning Electron Microscopy Experiment

The internal microstructure of the CGFB samples was examined using the Apreo S Hivac high-resolution SEM. The samples were soaked in alcohol and dried. After the surface of the samples was blown clean using an ear-washing ball, they were pasted on the sample table using a conductive adhesive. The internal structure of the samples was observed after the surface was sprayed with gold.

2.3.4. X-ray Diffraction Experiment

The crystal phase of the CGFB samples was analyzed using Rigaku Ultima IV XRD operating at a voltage of 40 kV and an emission current of 40 mA. The dried samples were ground using a mortar and screened using a 200-mesh sieve. During this, fragments with relatively few aggregates (coal gangue) were retained, and large particles such as aggregates (coal gangue) were removed. After the powder sample was tiled on the groove of the glass slide, it was placed into the instrument for testing.

2.3.5. Fourier-Transform Infrared Spectroscopy Experiment

The functional groups on the surface of the CGFB samples were measured using Nicolet iS5 FTIR. When testing samples, the samples were first ground with a mortar and then dried. During this, fragments with relatively few aggregates (coal gangue) were retained, and large particles such as aggregates (coal gangue) were removed. After mixing the samples with pure potassium bromide at a ratio of 1:10, grinding, and pressing, they were placed in the spectrometer and scanned 50 times to obtain the infrared spectrum.

3. Results

3.1. Uniaxial Compression Test Results

The strength of CGFB is an important index for evaluating coal mine safety conditions and backfill effects. In this test, the data are collected synchronously through a computer using a sampling interval of 10 ms. Table 2 shows the uniaxial compressive strength (UCS), peak strain, and elastic modulus of the samples. Figure 5 shows the uniaxial compressive stress–strain curves, while Figure 6 shows comparisons of the UCS, peak strain, and elastic modulus of CGFB samples under different amounts of kaolin.

Table 2. Uniaxial test results of CGFB samples.

Addition	Number	UCS (MPa)	Elastic Modulus (MPa)	Peak Strain (mm/mm)
0%	A-1	0.70	205.47	0.0061
	A-2	0.70	199.66	0.0068
	A-3	0.79	240.75	0.0064
	Average	0.73	215.29	0.0064
10%	B-1	0.58	211.20	0.0043
	B-2	0.80	183.47	0.0054
	B-3	0.67	182.44	0.0060
	Average	0.68	192.37	0.0053
20%	C-1	0.52	169.18	0.0048
	C-2	0.59	166.39	0.0043
	C-3	0.57	141.19	0.0049
	Average	0.56	158.92	0.0047
30%	D-1	0.31	156.33	0.0050
	D-2	0.39	123.17	0.0035
	D-3	0.46	179.88	0.0038
	Average	0.39	153.13	0.0041
40%	E-1	0.45	201.66	0.0025
	E-2	0.46	135.53	0.0045
	E-3	0.43	113.93	0.0037
	Average	0.45	150.37	0.0036
50%	F-1	0.35	72.36	0.0031
	F-2	0.36	90.85	0.0037
	F-3	0.38	61.91	0.0022
	Average	0.36	75.04	0.0030

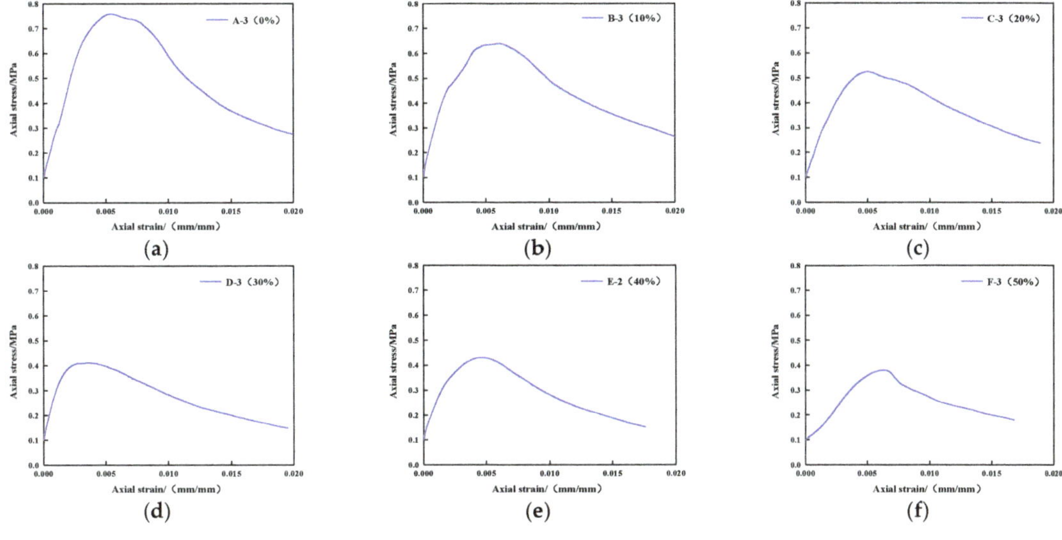

Figure 5. Stress–strain curves of CGFB samples. (**a**) A-3; (**b**) B-3; (**c**) C-3; (**d**) D-3; (**e**) E-2; (**f**) F-3.

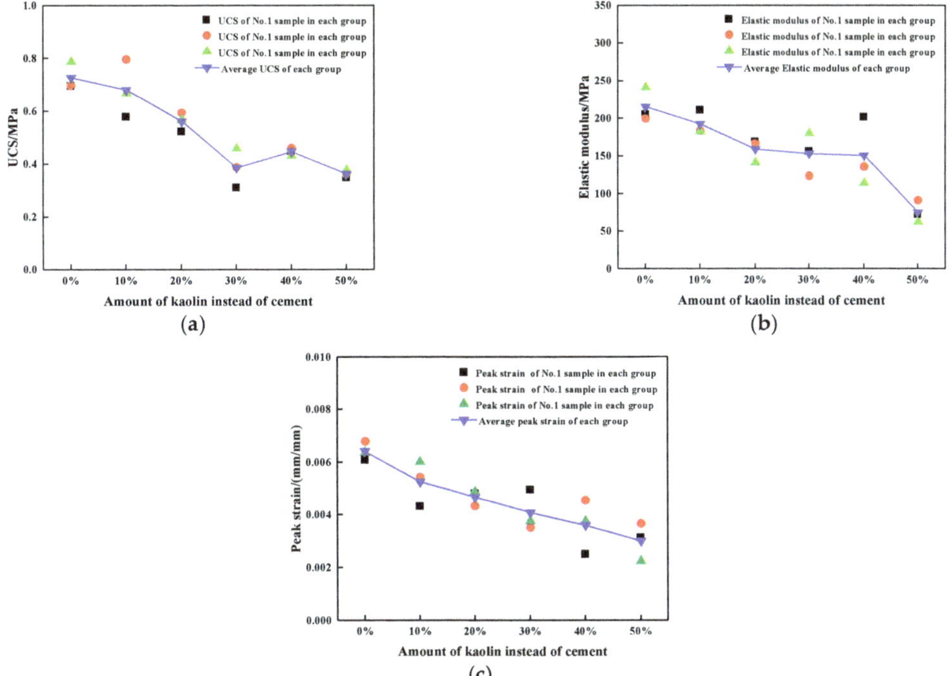

Figure 6. Comparisons of (**a**) UCS, (**b**) elastic modulus, and (**c**) peak strain of CGFB samples under different amounts of kaolin.

As seen from Figure 5, the stress–strain curves of the CGFB samples have the same shape, and they all pass through initial compaction, elastic deformation, plastic yield, and post-peak strain softening stages. However, the values of UCS, peak strain, and elastic modulus are different, illustrating that the kaolin affects the mechanical properties of CGFB samples. As shown in Figure 6, the UCS, peak strain, and elastic modulus of CGFB samples are affected by the amounts of kaolin replacing cement. CGFB samples with 0% kaolin instead of cement exhibit highest average UCS (0.73 MPa), elastic modulus (215.29 MPa), and peak strain (0.0064), while those with 50% kaolin exhibit lower UCS (0.36 MPa), elastic modulus (75.04 MPa), and peak strain (0.0030). The mechanical properties decrease with the increase of amounts of kaolin instead of cement. The higher the content of kaolin, the better the elasticity, and easier the deformation of CGFB. In addition, the amounts of kaolin instead of cement affect the post-peak strain softening stage of CGFB. The post-peak stress of CGFB decreases with time. The higher the content of kaolin, the smaller the slope of the CGFB curve, and stronger the plastic deformation ability.

In addition, in the Figure 6, it is found that the average values of UCS for CGFB samples decrease integrally with the amounts of kaolin instead of cement. Meanwhile, the values of UCS for the D-3 sample (30% CP) and group E samples (40% CP) show a slight increase. Previous investigations have shown that the strength of the CGFB samples is mainly determined by the cement hydration [30–34]. Generally, the stronger the cement hydration is, the larger the corresponding strength of the CGFB sample is. Normally, the activity of kaolin is lower than that of cement, and the incorporated kaolin only partially replaces cement for hydration reaction. The hydration caused by cement is lower than that of cement. With the increase of kaolin proportion, the amount of cement involved in the hydration reaction decreases. Therefore, the cement hydration effect is weakened, and the UCS of CGFB samples decreases. At the same time, the kaolin, which is not involved in hydration, mainly plays a role in filling the pores of CGFB samples. The average porosities of the failure or fracture surface of CGFB samples decrease with the kaolin content, which are analyzed in Section 3.3. Thus, the integrity of the CGFB samples is enhanced. Based on the above analyses, the strength of the CGFB samples containing kaolin may increase under these two mechanisms of the kaolin on mechanical properties for CGFB samples, but which is lower than that of the CGFB samples without the kaolin.

3.2. Macro Failure Characteristics

Figure 7 shows the macro-failure patterns of CGFB samples, which can be divided into tensile and shear failures, under different amounts of kaolin. The amounts of kaolin indeed affect the failure characteristics of CGFB samples as the failure degree increases with the increase of kaolin instead of cement. When the amount of kaolin is 50%, the highest amount of tensile cracks in CGFB samples is seen. The CGFB samples were gradually fractured along the diagonal direction with the increase of kaolin instead of cement. After the CGFB samples are fractured, the large tensile cracks on their surface increase, and most of them are distributed near the gangue particles. This shows that the decrease of cement content leads to the decrease of internal bonding degree. The ultimate failure characteristics of CGFB samples show tensile failure accompanied by local shear failure.

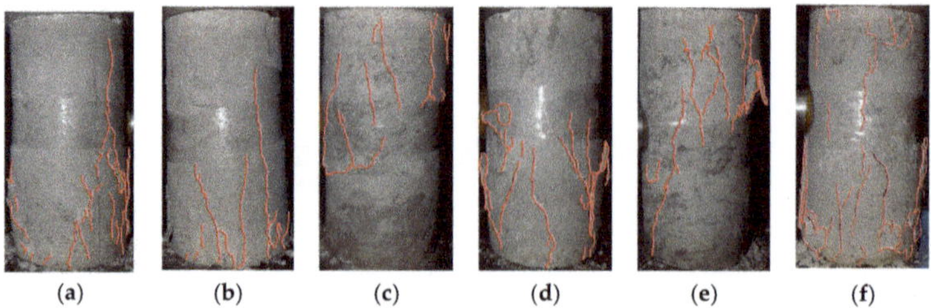

Figure 7. Failure characteristics of CGFB samples. (**a**) A-1; (**b**) B-2; (**c**) C-2; (**d**) D-3; (**e**) E-3; (**f**) F-3.

A single stress–strain curve cannot reflect the development process, but the failure degree of CGFB samples with different amounts of kaolin addition can be revealed in detail by analyzing the AE Ra and cumulative Ra value. The different characteristics of tensile failure and shear failure can also be understood in detail. The Ra value of the AE is the ratio of the rise time to amplitude, which is an important index for determining the fracture mode. Shiotani et al. [35] calculated the Ra value of rock under bending and shear tests and concluded that a low Ra value corresponds to a shear crack, while a high Ra value corresponds to a tensile crack.

In Figure 8, the relationship between axial stress and cumulative RA value over time during failure of CGFB samples with different kaolin contents is shown to reveal the failure characteristics. It can be seen that when the CGFB samples without kaolin fail, the initial Ra value is at a low level, and the cumulative Ra value increases slowly, which indicates the occurrence of shear failure. Near peak stress, the Ra value suddenly increases, and the cumulative Ra value increases rapidly. This phenomenon can be understood as the point when the CGFB samples without kaolin start getting compacted under the action of the vertical load, resulting in transverse tensile stress. As alluded to earlier, due to existence of gangue particles and a large number of holes, microcracks, and other defects in CGFB samples, their internal structure is under uneven stress, and the effective bearing area reduces, resulting in shear failure of mutual dislocation initially. With the gradual increase in vertical load, friction between a large number of holes and microcracks also increases, which inhibits mutual dislocation, and CGFB samples attain a stable state. As the load continues increasing and becomes higher than the compressive strength of CGFB samples, a crack begins to develop, and the sample finally fractures.

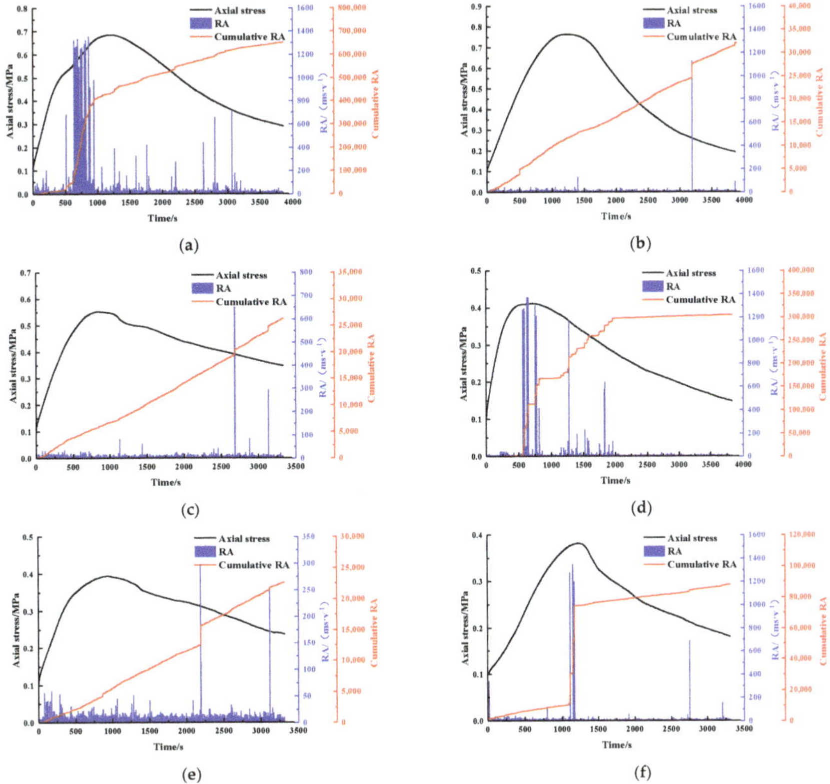

Figure 8. Acoustic emission (AE) characteristics of CGFB samples. (**a**) A-1; (**b**) B-2; (**c**) C-2; (**d**) D-3; (**e**) E-3; (**f**) F-3.

With the increase of the amount of kaolin instead of cement, the fluctuation of the Ra value of CGFB samples increases, and the difference of its cumulative Ra value decreases, which increases the number of tensile cracks in CGFB samples. This shows that a decrease in the cement content leads to a decrease in the degree of internal bonding. During the compression process, a large amount of elastic energy is stored in the gangue particles. With the increase in sample deformation, energy is released near the gangue particles and causes chain failure of the surrounding structure, resulting in tensile failure accompanied by local shear failure of the sample.

3.3. Microstructure Characteristics

The microstructure and morphology of the fracture surface of CGFB samples are analyzed using SEM and are shown in Figure 9. It can be seen that the microstructure and morphology of fracture surfaces of different CGFB samples are different. Because of the disappearance of the coating layer on the surface of CGFB samples, the rate of hydration reaction increases, and a large amount of flocculated calcium-silicate-hydrate (C-S-H) gel forms in the main body. Ettringite (AFt) having a needle-like structure is mostly distributed in pores. These crystals are closely connected, which improves the strength of the sample. When the amount of kaolin is 10%, more hydration products (AFt and C-S-H gel) are observed. The internal structure of the samples is compact, and the number of pores is small. When the amount of kaolin is more than 10%, the number of hydration products decreases, and that of irregular particles increases. At 50% kaolin content, fewer hydration products are observed, but the internal structure is relatively dense, which indicates that kaolin mainly fills the pores in higher quantities.

Figure 9. SEM image of fracture surface of CGFB samples. (**a**) 0% CP; (**b**) 10% CP; (**c**) 20% CP; (**d**) 30% CP; (**e**) 40% CP; (**f**) 50% CP.

PCAS is a professional software tool for the identification and quantitative analysis of pore systems and fracture systems [36]. It can automatically identify all types of pores and fractures in images and obtain all geometric and statistical parameters. Concerning pore recognition, it can import various pore images, remove clutter automatically, segment and recognize pores automatically through binarization, output the geometric and statistical parameters, and display the result vector image and rose diagram. It can also display various geometric parameters of all pores in the data table, including the number of pores, area, length, width, directivity, and shape coefficient, and obtain statistical parameters such as region percentage (porosity), average form factor, probability entropy, fractal dimension, and sorting coefficient [37]. In this study, PCAS is used to quantitatively analyze the micropore structure of CGFB samples. The pore map after PCAS processing is shown in Figure 10, where the black zone represents the non-porous area, and the colored zone represents pores where the software automatically recognizes and artificially corrects some recognition errors and defects. For different pores (unconnected pores), the PCAS uses a different color mark, while the same color mark is used for connected pores [38]. All CGFB samples are observed and analyzed in this manner. The minimum diameter of the closed pore in PCAS is 2, and the minimum pore area is 50. The region percentages (porosities) of the samples are listed in Table 3.

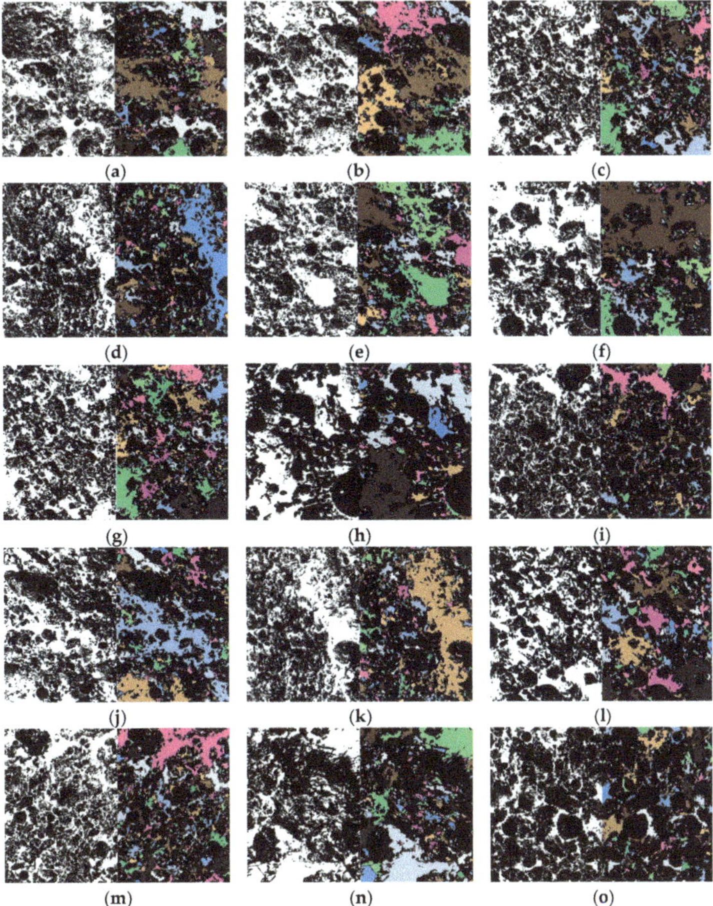

Figure 10. *Cont.*

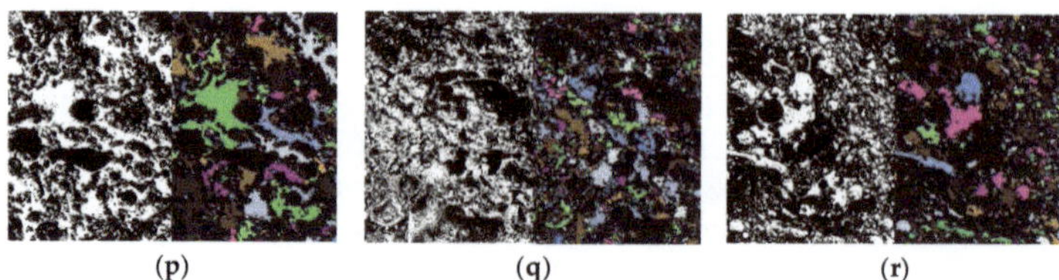

(p)   (q)   (r)

Figure 10. Binarization of CGFB microscopic pores. (**a**) A-1; (**b**) A-2; (**c**) A-3; (**d**) B-1; (**e**) B-2; (**f**) B-3; (**g**) C-1; (**h**) C-2; (**i**) C-3; (**j**) D-1; (**k**) D-2; (**l**) D-3; (**m**) E-1; (**n**) E-2; (**o**) E-3; (**p**) F-1; (**q**) F-2; (**r**) F-3.

Table 3. Region percentage (porosity) of CGFB samples.

Addition	Number	Region Percentage (Porosity)
0%	A-1	43.26%
	A-2	49.45%
	A-3	41.26%
	Average	44.66%
10%	B-1	33.52%
	B-2	45.21%
	B-3	46.56%
	Average	41.76%
20%	C-1	41.27%
	C-2	39.64%
	C-3	39.94%
	Average	40.28%
30%	D-1	39.08%
	D-2	39.54%
	D-3	38.47%
	Average	39.03%
40%	E-1	37.08%
	E-2	36.04%
	E-3	36.05%
	Average	36.39%
50%	F-1	35.07%
	F-2	34.84%
	F-3	32.77%
	Average	34.23%

Many studies [39–42] have shown that porosity has a significant relationship with UCS. Table 3 shows the porosity of CGFB samples with different amounts of kaolin instead of cement. The average porosity of each group of CGFB is compared with the average UCS, and their relationship under different amounts of kaolin instead of cement is shown in Figure 11.

Generally speaking, when the porosity of CGFB samples decreases, the UCS also decreases. However, in Figure 11, an opposite trend is seen. The reason is that the hydration is the main factor of kaolin affecting the strength of CGFB samples. The activity of kaolin is lower than that of cement, but with the increase of kaolin content, the hydration of cement decreases, so the UCS of CGFB samples decreases. However, kaolin also fills the internal pores of CGFB samples, which improves their integrity and reduces the corresponding porosity. In summary, the porosity of CGFB samples decreases with the increase of kaolin instead of cement, and the UCS gradually decreases.

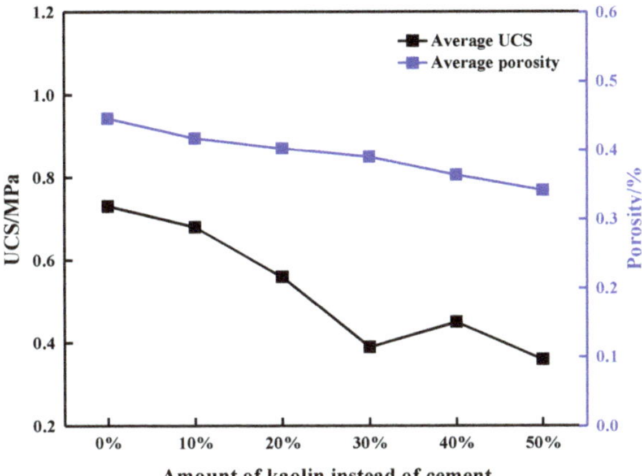

Figure 11. Comparison of the relationship between UCS and porosity of CGFB with different amount of kaolin instead of cement.

4. Discussion

The goal of coal mine filling is to meet the needs of coal mine production safety at the lowest possible cost. Therefore, the selection of appropriate CGFB samples material is related not only to the filling cost but also the safety and stability of the goal. After adding kaolin into CGFB samples, the sample will have effects of morphology, micro-aggregate, and activity of the kaolin admixture. The morphological effects of kaolin addition are mainly reflected in the influence of kaolin particle size, shape, and other factors on the performance of CGFB samples. The micro-aggregate effects are that the particles are relatively fine and can be evenly dispersed in the gangue aggregate and flocculation structure, filling the internal pores, which helps improve the internal uniformity of CGFB samples. Kaolin is rich in active SiO_2 and Al_2O_3. The essence of its pozzolanic activity is that SiO_2 and Al_2O_3 are excited in alkaline environments. The content of soluble active components in kaolin is very low; therefore, the degree of pozzolanic reaction of kaolin is low initially.

To study the influence of different amounts of kaolin content on the hydration products of CGFB samples, phases of CGFB samples cured for 28 days are analyzed via XRD, and the diffraction patterns are shown in Figure 12. Similar diffraction patterns of CGFB samples are seen, but the intensity of the peaks is different from that of the hydrated products, which form $Ca(OH)_2$, C-S-H gel, AFt, etc. When the amount of kaolin is less than 10%, the peak of $Ca(OH)_2$ becomes weaker, the SiO_2 peak becomes stronger, and the C-S-H gel diffraction peak becomes stronger with an increase in the amount of kaolin instead of cement. This indicates that some active substances in kaolin consume $Ca(OH)_2$ to participate in the two hydration reactions. When the amounts of kaolin instead of cement are greater than 10%, the peak value of SiO_2 continues to increase, and the diffraction peaks of $Ca(OH)_2$, AFt, and C-S-H become weak with an increase in the amount of kaolin content. This shows that after a certain point of kaolin addition, reduction in cement content per unit volume will negatively affect the secondary hydration of kaolin.

To further investigate the effects of kaolin on CGFB samples, the chemical structures of the prepared samples are characterized via FTIR spectroscopy. Figure 13 shows the infrared spectrum of CGFB samples, which shows that FTIR spectra of CGFB samples are similar. The broad absorption band at wavenumbers of 2976.99–3592.77 cm^{-1} charac-terizes the stretching vibration of Al-OH in the $[AlO_4]$ tetrahedron, and the absorption peak of 1588.25 to 1784.38 cm^{-1} represents the bending vibration of H_2O in C-S-H [43]. Absorption peaks at 1349.18–1588.25 cm^{-1} represent the O-C-O asymmetric stretching vibration of carbonate, indicating that the CGFB samples experienced slight carbonization during the characterization process [44]. Absorption peaks of 936.78–1349.18 cm^{-1} corresponds to the asymmetric stretching vibration of Si-O in tetrahedron [45], while those at 820.31–886.33 cm^{-1} reflects the existence of [Al(Fe)-O], which means that some Al-OH in AFt is replaced by Fe-OH [46].

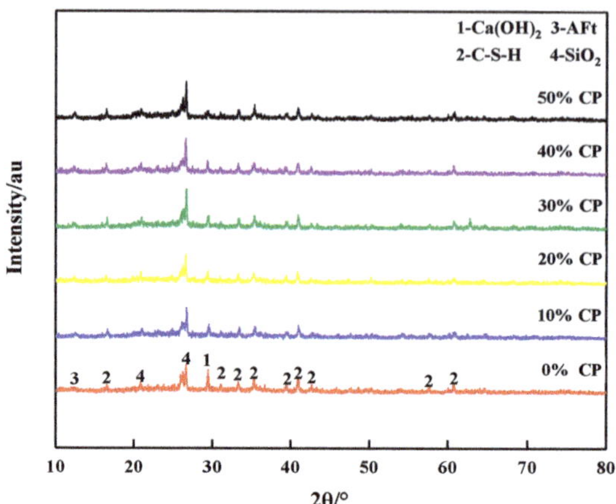

Figure 12. XRD patterns of CGFB samples.

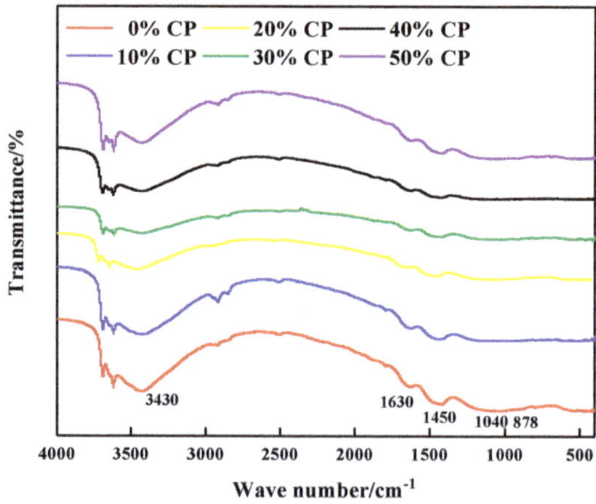

Figure 13. FTIR spectrum of CGFB samples.

Figure 14 gives the pozzolanic reaction of kaolin in the CGFB sample. Pore water rich in Ca^{2+}, AlO$_2^-$, and SiO$_3^{2-}$ first infiltrate CGFB, and the cement particles in CGFB samples hydrate to form Ca(OH)$_2$ and other products in the liquid phase. The hydration reaction of cement is as follows:

$$C_3S + nH \rightarrow C\text{-}S\text{-}H + (3\text{-}x)\, CH \tag{1}$$

$$C_2S + mH \rightarrow C\text{-}S\text{-}H + (2\text{-}x)\, CH \tag{2}$$

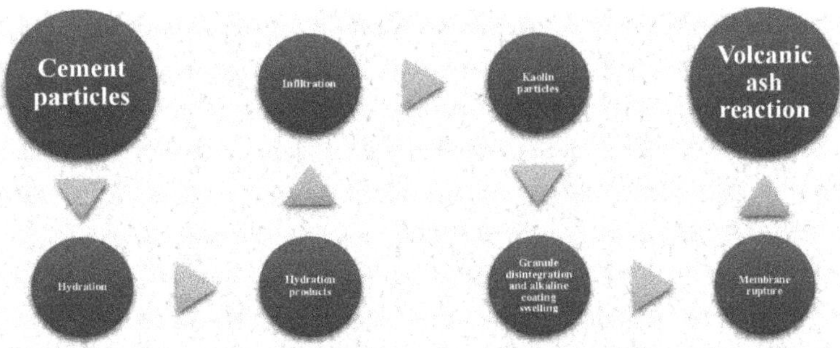

Figure 14. Pozzolanic reaction process of kaolin in the CGFB sample.

The hydration product will infiltrate the kaolin particles. During slurry condensation, the hydration products in the matrix crystallize, and the CGFB samples material have a certain strength through ionic bonds and intermolecular forces. Now, alkaline inclusions gradually form on the surface of the kaolin particles. With an increase in the curing time from 0 to 28 days, kaolin particles are gradually eroded by alkaline inclusions, resulting in chemical changes and development of the active state. Because of the different ion concentrations inside and outside the coating, ion penetration expands the coating. When the pressure of the coating reaches its limit, the active component reacts with the ions to form C-S-H and other products. The formula for secondary hydration of kaolin is as follows:

$$SiO_2 + m_1 Ca(OH)_2 + xH2O \rightarrow m_1 CaO \cdot SiO_2 \cdot xH_2O \tag{3}$$

$$Al_2O_3 + m_2 Ca(OH)_2 + yH2O \rightarrow m_2 CaO \cdot Al_2O_3 \cdot yH_2O \tag{4}$$

The test results show that the strength of CGFB samples decreases with an increase in kaolin content. When the amount of kaolin instead of cement is 10%, the contribution of secondary hydration of some active components in kaolin to CGFB strength is lower than that of normal hydration of cement, but kaolin has a certain filling effect. Therefore, in the study, the strength of CGFB samples with kaolin sample decreased, but there is no significant difference compared with those of CGFB samples without kaolin, which is approximately 0.7 MPa. When the amount of kaolin is greater than 10%, the activity of kaolin is lower than that of cement, which mainly plays the role of filling. The decrease in cement content directly leads to a decrease in the hydration products in CGFB samples, a decrease in the cohesive force, and a consequent decrease in CGFB strength.

Therefore, considering requirements of CGFB strength for coal mine production safety, this study recommends replacing 10% cement in CGFB samples by kaolin as the most suitable option among those tested. That is to say, when the kaolin content is 10%, it can play a similar role as cement in CGFB samples and meet the mine-filling strength requirements. This paper mainly discusses the influence of kaolin on mechanical properties and microstructure of CGFB; however, carrying out systematic research on the durability and engineering applications of CGFB samples with kaolin is also necessary.

5. Conclusions

(1) The stress–strain curves of the CGFB samples are essentially the same, and they all pass through initial compaction, elastic deformation, plastic yield, and post-peak strain softening stages. The uniaxial compressive strength, peak strain, and elastic modulus decrease with the kaolin content. The average uniaxial compressive strength, elastic modulus, and peak strain of CGFB samples with 10% amount of kaolin are 0.68, 192.37, and 0.0053 MPa, respectively, which are close to those of CGFB samples with no kaolin.

(2) The kaolin content affects the failure characteristics of CGFB samples, which show tensile failure accompanied by local shear failure, and the failure degree increases

with the kaolin content. The fluctuation of Ra value of CGFB samples increases, and the difference of its cumulative Ra value decreases with the kaolin content, which increases tensile cracks in CGFB samples. The porosity of the fracture surface shows a decreasing trend as a whole. The reason is that the hydration is the main factor of kaolin affects the strength of CGFB samples, and the activity of kaolin is lower than that of cement. With the increase of kaolin content, the hydration of cement decreases, so the UCS of CGFB samples decreases. However, kaolin also fills the internal pores of CGFB samples, which improves their integrity and reduces the corresponding porosity. In summary, the porosity of CGFB samples decreases with the increase of kaolin, and the UCS gradually decreases.

(3) When the amount of kaolin is 10%, the internal structure of the CGFB sample is more compact, and the number of pores is less. When it is more than 10%, with an increase in the kaolin content, the decrease in cement content per unit volume leads to a decrease in the number of AFt and C-S-H gel, the peak of the SiO_2 diffraction peak becomes stronger, the C-S-H diffraction peak becomes weaker, and the number of irregular particles increases. As the average uniaxial compressive strength, elastic modulus, and peak strain of CGFB samples with 10% amount of kaolin are close to those of CGFB samples with no kaolin, replacing 10% cement in CGFB samples by kaolin is the most suitable option recommended.

Author Contributions: Conceptualization, D.Y.; methodology, D.Y.; software, F.L.; validation, C.Z., F.W. and N.J.; data curation, Z.Z.; writing—original draft preparation, F.L.; writing—review and editing, F.L. and D.Y.; funding acquisition, D.Y. All authors have read and agreed to the published version of the manuscript.

Funding: This research was funded by the National Science Foundation of China, grant number 51904167 and 52074169; the Taishan Scholars Project; the Taishan Scholar Talent Team Support Plan for Advantaged and Unique Discipline Areas; the SDUST Research Fund; and the Open Research Fund for the Key Laboratory of Safety and High-efficiency Coal Mining, grant number JYBSYS2019201.

Data Availability Statement: The data are available and explained in this article; readers can access the data supporting the conclusions of this study.

Conflicts of Interest: The authors declare no conflict of interest.

References

1. Miao, X.X.; Ju, F.; Huang, Y.L.; Guo, G.L. New progress and prospect of backfill mining theory and technology. *J. China Univ. Min. Technol.* **2015**, *44*, 391–399, 429.
2. Liu, Y.; Guo, Y.L.; Li, H.; Wang, H.Y. Experimental study on the influence of construction waste recycled aggregate on the transportation performance of mine CGFB. *J. Shandong Univ. Sci. Technol. Nat. Sci. Ed.* **2020**, *39*, 59–65.
3. Sun, X.K. Development status and prospect of green filling mining in mines. *Coal Sci. Technol.* **2020**, *48*, 48–55.
4. Liu, J.G.; Li, X.W.; He, T. Application status and development of filling mining in China. *J. China Coal Soc.* **2020**, *45*, 141–150.
5. Mo, Z.Y.; Liu, Y.L.; Wang, D.G.; Li, Z.J.; Bai, L.G. Research progress on mechanical properties of metakaolin cement based materials. *Chin. Ceram. Soc.* **2018**, *37*, 911–917.
6. Cao, Y.D.; Li, Y.X.; Zhang, J.S.; Cao, Z.; Sun, C.B. Effects of fineness and calcination temperature on pozzolanic activity and microstructure of coal gangue. *Chin. Ceram. Soc.* **2017**, *45*, 117–122.
7. Cheng, H.L.; Yang, F.H.; Ma, B.G.; Zhang, J.; Hao, L.W.; Liu, G.Q. Composite activation of high alumina gangue and analysis of its pozzolanic effect. *J. Build. Mater.* **2016**, *19*, 248–254.
8. Lin, C.; Li, J.X.; Chen, P.X.; Chen, G. Study on the influence of metakaolin on the compressive strength of concrete. *Guangdong Arch. Civ. Eng.* **2019**, *26*, 70–72.
9. Hou, L.Y.; Yang, A.R. Study on the influence of mineral admixtures on cement hydration properties. *Water Res. Hydropower Eng.* **2020**, *51*, 198–204.
10. Pang, J.Y.; Chen, X.P. Mechanical properties test of concrete with high active mineral admixture. *Chin. Ceram. Soc.* **2020**, *39*, 3143–3151.
11. Xi, Y.S. Effect of metakaolin admixture on workability and early mechanical properties of cement mortar. *China Concr.* **2019**, *8*, 72–76.
12. Jiang, G.; Rong, Z.D.; Sun, W. Effect of metakaolin on properties of high performance cement mortar. *J. Southeast Univ. Nat. Sci. Ed.* **2015**, *45*, 121–125.
13. Li, F.H.; Zhang, G.B.; Zhou, H.Y.; Zhou, S.; Li, G.H. Inhibitory effect of high activity metakaolin and fly ash on alkali aggregate reaction. *J. Build. Mater.* **2017**, *20*, 876–880.
14. Dong, M.; Ma, J.D.; Li, Y.F. Effect of metakaolin admixture on strength of cement mortar. *J. Mater. Sci. Eng.* **2020**, *38*, 295–300.

15. Liu, Y.Y. Application and Theoretical Study of Coal Series Kaolin Activation and Cement Admixture. Ph.D. Thesis, Wuhan University of Technology, Wuhan, China, 2018.
16. Wang, H. Process Mineralogy and Beneficiation Test of Sandy Kaolin. Master's Thesis, Wuhan University of Technology, Wuhan, China, 2013.
17. Yu, W. Mechanism of Metakaolin on Hydration Products of Cement Paste. Master's Thesis, Wuhan University of Technology, Wuhan, China, 2013.
18. Ebenezer, A.; Emmanuel, N.; Benjamin, A.T.; Stephen, K.A.; George, N.B.; Joanna, A.M.H.; Michael, O.P. Synthesis and Characterization of Modified Kaolin-Bentonite Composites for Enhanced Fluoride Removal from Drinking Water. *Adv. Mater. Sci. Eng.* **2021**, *2021*, 6679422.
19. Chen, M.N.; Chen, X.L.; Zhang, C.Y.; Cui, B.Z.; Li, Z.W.; Zhao, D.Y.; Wang, Z. Kaolin-Enhanced Superabsorbent Composites: Synthesis, Characterization and Swelling Behaviors. *Polymers* **2021**, *13*, 1204. [CrossRef]
20. Partha, D.; Tadikonda, V.B. Kaolin based protective barrier in municipal landfills against adverse chemo-mechanical loadings. *Sci. Rep.* **2021**, *11*, 1–12.
21. Chen, M.C.; Fang, W.; Xu, K.C. Effect of ceramic admixtures on performance of cement mortar. *Chin. Ceram. Soc.* **2015**, *34*, 793–798.
22. Wang, Z.Y.; Yao, S.Y. Structure and properties of animal bone ash doped Na_2O-CaO-SiO_2 Wollastonite Glass ceramics. *J. Shandong Univ. Sci. Technol. Nat. Sci. Ed.* **2020**, *39*, 48–55.
23. Zhang, H.W.; Elsworth, D.; Wan, Z.J. Failure response of composite rock-coal samples. *Geomech. Geophys. Geo-Energy Geo-Resour.* **2018**, *4*, 175–192. [CrossRef]
24. Yin, D.W.; Chen, S.J.; Liu, X.Q.; Ma, H.F. Effect of joint angle in coal on failure mechanical behavior of roof rock-coal combined body. *Q. J. Eng. Geol. Hydrogeol.* **2018**, *51*, 202–209. [CrossRef]
25. Yin, D.W.; Chen, S.J.; Chen, B.; Liu, X.Q.; Ma, H. Strength and failure characteristics of the rock-coal combined body with single joint in coal. *Geomech. Eng.* **2018**, *15*, 1113–1124.
26. Lu, Y.W.; Sun, C.P.; Shen, B.T. Experimental study on damage evolution and crack propagation characteristics of sandstone under combined stress. *J. Shandong Univ. Sci. Technol. Nat. Sci. Ed.* **2020**, *39*, 37–45.
27. Wang, Q.; Jiang, Z.H.; Jiang, B.; Gao, H.K.; Huang, Y.B.; Zhang, P. Research on an automatic roadway formation method in deep mining areas by roof cutting with high-strength bolt-grouting. *Int. J. Rock Mech. Min.* **2020**, *128*, 104264. [CrossRef]
28. Yin, S.J.; Chen, S.J.; Ge, Y.; Liu, R. Mechanical properties of rock–coal bi-material samples with different lithologies under uniaxial loading. *J. Mater. Res. Technol.* **2021**, *10*, 322–338. [CrossRef]
29. ASTM E569/E569M-13. *Standard Practice for Acoustic Emission Monitoring of Structures during Controlled Stimulation*; ASTM International: West Conshohocken, PA, USA, 2013; Available online: www.astm.org (accessed on 10 March 2021).
30. Sukpyo, K.; Hyeju, K.; Byoungky, L. Hydration Properties of Cement with Liquefied Red Mud Neutralized by Nitric Acid. *Materials* **2021**, *14*, 2641. [CrossRef]
31. Geuntae, H.; Sangwoo, O.; Seongcheol, C.; Won-Jong, C.; Young-Jin, K.; Chiwon, S. Correlation between the Compressive Strength and Ultrasonic Pulse Velocity of Cement Mortars Blended with Silica Fume: An Analysis of Microstructure and Hydration Kinetics. *Materials* **2021**, *14*, 2476. [CrossRef]
32. Cao, K.; Wang, L.; Xu, Y.; Shen, W.F.; Wang, H. The Hydration and Compressive Strength of Cement Mortar Prepared by Calcium Acetate Solution. *Adv. Civ. Eng.* **2021**, *2021*, 8817725. [CrossRef]
33. Tragazikis, I.K.; Kordatou, T.Z.; Exarchos, D.A.; Dalla, P.T.; Matikas, T.E. Monitoring the Hydration Process in Carbon Nanotube Reinforced Cement-Based Composites Using Nonlinear Elastic Waves. *Appl. Sci.* **2021**, *11*, 1720. [CrossRef]
34. Zhang, J.; Ke, G.J.; Liu, Y.Z. Early Hydration Heat of Calcium Sulfoaluminate Cement with Influences of Supplementary Cementitious Materials and Water to Binder Ratio. *Materials* **2021**, *14*, 642. [CrossRef]
35. Shiotani, T.; Ohtsu, M.; Ikeda, K. Detection and evaluation of AE waves due to rock deformation. *Constr. Build. Mater.* **2001**, *15*, 235–246. [CrossRef]
36. Liu, C.; Shi, B.; Zhou, J.; Tang, C. Qunication and characterization of microporosity by image processing, geometric measurement and statistical methods: Application on SEM images of clay materials. *Appl. Clay Sci.* **2011**, *54*, 97–106. [CrossRef]
37. Liu, C.; Tang, C.; Shi, B.; Suo, W. Automatic quantification of crack patterns by image processing. *Comput. Geosci.* **2013**, *57*, 77–80. [CrossRef]
38. Liu, C.; Xu, Q.; Shi, B.; Gu, Y.F. Digital image recognition method of rock particle and pore system and its application. *Geotech. Eng.* **2018**, *40*, 925–931.
39. Urfels, S.; Bauer, J.; Vogel, H. General method of specimen preparation for the quantitative microscopic characterization of nano- and microparticles. *Mater. Charact.* **2019**, *153*, 60–68. [CrossRef]
40. Farnama, Y.; Geikerb, M.R.; Bentzc, D.; Weiss, J. Acoustic emission waveform characterization of crack origin and mode in fractured and ASR damaged concrete. *Cem. Concr. Compos.* **2015**, *60*, 135–145. [CrossRef]
41. Zingg, L.; Briffaut, M.; Baroth, J.; Malecot, Y. Influence of cement matrix porosity on the triaxial behaviour of concrete. *Cem. Concr. Res.* **2016**, *80*, 52–59. [CrossRef]
42. Chen, G.; Xu, P.; Mi, G.; Zhu, J. Compressive strength and cracking of composite concrete in hot-humid environments based on microscopic quantitative analysis. *Constr. Build. Mater.* **2019**, *225*, 441–451. [CrossRef]

43. Chen, J.L.; Zhang, N.; Li, H.; Zhao, X.B.; Liu, X.M. Study on hydration characteristics of red mud based paste like filling materials. *Chin. J. Chem. Eng.* **2017**, *39*, 1640–1646.
44. Li, Y.C.; Min, X.B.; Ke, Y.; Chai, L.Y.; Shi, M.Q.; Tang, C.J.; Wang, Q.W.; Liang, Y.J.; Lei, J.; Liu, D.G. Utilization of red mud and Pb/Zn smelter waste for the synthesis of a red mud-based cementitious material. *J. Hazard. Mater.* **2018**, *344*, 343–349. [CrossRef] [PubMed]
45. Andini, S.; Cioffi, R.; Colangelo, F. Coal fly ash as raw material for the manufacture of geopolymer-based products. *Waste Manag.* **2008**, *28*, 416–423. [CrossRef] [PubMed]
46. Singh, M.; Upadhayay, S.N.; Prasad, P.M. Preparation of iron rich cements using red mud. *Cem. Concr. Res.* **1997**, *27*, 1037–1046. [CrossRef]

Article

Seismic Response Analysis of Tunnel through Fault Considering Dynamic Interaction between Rock Mass and Fault

Guoqing Liu [1,*], Yanhong Zhang [1], Junqing Ren [2] and Ming Xiao [2]

[1] State Key Laboratory of Simulation and Regulation of Water Cycle in River Basin, China Institute of Water Resources and Hydropower Research, Beijing 100048, China; zhangyh@iwhr.com

[2] State Key Laboratory of Water Resources and Hydropower Engineering Science, Wuhan University, Wuhan 430072, China; junqingrw@whu.edu.cn (J.R.); mxiao@whu.edu.cn (M.X.)

* Correspondence: liugq@iwhr.com

Abstract: In order to explore the influences of fault dislocations on tunnel stability under seismic action, a nonlinear dynamic simulation method for the rock–fault contact system is proposed. First, considering the deterioration effect of seismic action on the ultimate bearing load of the contact interface between rock mass and fault, a mathematical model is established reflecting the seismic deterioration laws of the contact interface. Then, based on the traditional point-to-point contact type in a geometric mesh, a point-to-surface contact type is also considered, and an improved dynamic contact force method is established, which considers the large sliding characteristics of the contact interface. According to the proposed method, a dynamic finite element calculation for the flow of the rock–fault contact system is designed, and the accuracy of the method is verified by taking a sliding elastic block as an example. Finally, a three-dimensional (3D) calculation model for a deep tunnel through a normal fault is built, and the nonlinear seismic damage characteristics of the tunnel under horizontal seismic action are studied. The results indicate that the relative dislocation between the rock mass and the fault is the main factor that results in lining damage and destruction. The seismic calculation results for the tunnel considering the dynamic interaction between the rock mass and the fault can more objectively reflect the seismic response characteristics of practical engineering. In addition, the influences of different fault thicknesses and dip angles on the seismic response of the tunnel are discussed. This work provides effective technical support for seismic fortification in a tunnel through fault.

Keywords: tunnel through fault; concrete lining structure; fault dislocation; seismic deterioration effect; dynamic contact force; seismic damage analysis

Citation: Liu, G.; Zhang, Y.; Ren, J.; Xiao, M. Seismic Response Analysis of Tunnel through Fault Considering Dynamic Interaction between Rock Mass and Fault. *Energies* **2021**, *14*, 6700. https://doi.org/10.3390/en14206700

Academic Editor: Nikolaos Koukouzas

Received: 8 September 2021
Accepted: 12 October 2021
Published: 15 October 2021

Publisher's Note: MDPI stays neutral with regard to jurisdictional claims in published maps and institutional affiliations.

1. Introduction

A large number of tunnel projects in China that are under construction or have already been built, are located in earthquake-prone areas, especially in the southwest. The seismic response of tunnels in areas of high earthquake intensity is a problem that must be faced in current tunnel construction, especially for tunnels through fault fracture zones [1–3]. The "5.12" Wenchuan earthquake indicated that tunnels with good engineering geological conditions show good seismic performance, while tunnels with complex geological conditions, including major changes of strata or poor rock properties, are more vulnerable to seismic damage. Therefore, the fault fracture zone is the main area where the tunnel's seismic damage is concentrated [4–6]. Engineering practice indicates that once an earthquake occurs, relative dislocation is easily produced between the rock mass and the fault, which can lead to irreparable damage to the tunnel and can affect its normal operation. This also makes rescue work in the earthquake area much more difficult. Therefore, it is necessary to study the seismic damage characteristics of the tunnel through fault, which will have great practical engineering value and social economic benefit.

At present, model tests and numerical simulations are the main methods adopted to study the dynamic response laws of the tunnel through fault under seismic action. In a model test, Fang et al. [7] conducted a large-scale shaking table test study of a complex tunnel through fault constructed in Tibet and analyzed the propagation laws of seismic waves and the failure modes of the tunnel. Liu and Gao [8] studied the dynamic failure characteristics of the tunnel through fault compared with ordinary tunnels, using shaking table tests, and they analyzed the development process for lining cracks and the distribution laws of dynamic earth pressure. Fan et al. [9] designed a 3D sliding device representing an active fault and conducted a shaking table test to investigate the seismic performance of a tunnel under normal fault sliding. Although many scholars have achieved a great deal in shaking table test research on the seismic response behavior of the tunnel through fault, the development and application of model tests are still limited by many factors. For example, the engineering geological environment and the boundary conditions are difficult to grasp accurately, and the test cost is very high.

Relatively speaking, the prospects for the application of numerical simulations are more promising than those for model tests, due to their low cost and high efficiency. Li et al. [10] analyzed the displacement difference and plastic zone of a railway tunnel through fault under seismic excitation, using a solid element and a structural plane element to simulate the fault. Yang et al. [11] studied the changing process of stress and strain in a tunnel under the joint action of an earthquake and a fault, using the discrete element method. Ardeshiri et al. [12] conducted a nonlinear dynamic analysis of a cavern–fault system, using the hybrid finite difference–discrete element method. The key to the numerical simulation of seismic damage in the tunnel through fault lies in establishing a reasonable analysis method for the dynamic interaction between the rock mass and the fault. However, most current studies assume that the rock mass and the fault constitute a continuous system, which ignores the influence of discontinuous fault dislocations on the dynamic response of the lining structure. In fact, the dynamic contact force method proposed by Liu and Sharan [13] provides a good method for solving this kind of problem that does not need iteration and is easy to combine with the explicit central difference method [14,15]. However, the traditional dynamic contact force method assumes a small sliding displacement, which is clearly not perfect when simulating large dislocations between rock masses and faults.

In this paper, an improved dynamic contact force method is established that considers the seismic deterioration effect and the large sliding characteristics of the contact interface between the rock mass and the fault. A calculation formula for the vibration deterioration coefficient of the contact interface is obtained, taking into account the deterioration effect of seismic action on the ultimate bearing load of the contact interface. Based on the traditional point-to-point contact type, this method also takes the point-to-surface contact type into account, and the solving strategies for the contact force and methods of judging the contact state in the two cases are expounded. Then, the calculation flow of the improved dynamic contact force method suitable for the tunnel contact system is designed. The simulation results for the engineering case indicate that the proposed method reflects the nonlinear seismic damage characteristics of the tunnel through fault reasonably well.

2. Seismic Deterioration Effect of Contact Interface

The dynamic deterioration effect of the rock structural plane under seismic action has been confirmed by a large number of rock dynamic tests [16–18]. Through cyclic shear tests on the structural plane for iron-cemented fine sandstone from the Wenchuan earthquake area, Ni et al. [19] obtained a quantitative expression reflecting the vibration deterioration effect of the structural plane. By introducing the vibration deterioration model and making some appropriate modifications, this paper established a mathematical model suitable for studying the deterioration laws of the ultimate bearing load of the contact interface between the rock mass and the fault under seismic action.

2.1. Influencing Factors for the Seismic Deterioration of Contact Interface

The deterioration effect of seismic action on the contact interface is a complex, dynamically changing process, and the deterioration degree can be quantitatively described by the vibration deterioration coefficient D^t (where t is the time). Summarizing previous research results [20–22], it is found that the influences of dynamic cyclic action on the shear properties of the structural plane mainly include the factors of relative velocity and vibration wear. This paper adopts the influence coefficient of relative velocity γ^t and the influence coefficient of vibration wear η^t to quantitatively describe the influences of relative velocity and cyclic action on the ultimate bearing load of the contact interface. Assuming that the relative velocity and the vibration wear are two independent factors throughout the whole duration of the earthquake process, the vibration deterioration coefficient can be expressed as

$$D^t = \gamma^t \eta^t \tag{1}$$

2.2. Calculation Formula for the Vibration Deterioration Coefficient

Li et al. [23] conducted rock shear tests under different shear velocities, normal stresses, and undulating angles, using artificial concrete simulation samples. It was found that as the shear deformation velocity increased, the shear strength of the rock joint decreased, and the decreasing rate also decreased gradually. The attenuation laws of the joint strength presented negative exponential changing characteristics, so the influence coefficient of the relative velocity could be quantitatively described using a negative exponential model:

$$\gamma^t = \alpha \left(\left| \Delta \dot{u}^t \right| + \beta \right)^{-\lambda} \tag{2}$$

where α, β, and λ are undetermined coefficients and $\Delta \dot{u}^t$ is the relative shear velocity in mm/s.

Experimental research [19] indicates that as the number of shear cycles increases, the shear strength of the structural plane decreases, and decreases rapidly in the initial stage. After several shear cycles, the strength value remains unchanged, which can also be described by negative exponential attenuation laws. Therefore, the influence coefficient of the vibration wear can be expressed as

$$\eta^t = P_0 + (1 - P_0) \exp\left(-\xi \frac{t}{t_{\mathrm{w}}} \right) \tag{3}$$

where ξ is an undetermined coefficient, t_{w} is the whole duration of the earthquake, and P_0 is a convergence constant.

Combining Equations (1)–(3), the calculation formula for the vibration deterioration coefficient of the contact interface can be obtained:

$$D^t = \left[\alpha \left(\left| \Delta \dot{u}^t \right| + \beta \right)^{-\lambda} \right] \left[P_0 + (1 - P_0) \exp\left(-\xi \frac{t}{t_{\mathrm{w}}} \right) \right] \tag{4}$$

The undetermined coefficients α, β, λ, P_0, and ξ in Equation (4) should be determined according to the shear tests of rock materials in practical engineering. The relative shear velocity $\Delta \dot{u}^t$ of the contact interface is obtained by the time-step integration of finite elements, so that the vibration deterioration coefficient D^t can be solved explicitly in each time step for seismic loading.

3. Contact Conditions and Contact States

3.1. Contact Conditions

Figure 1 shows a typical contact problem. Ω_1 and Ω_2 are two objects, whose boundaries are Γ_1 and Γ_2, respectively. S_u and S_σ are the given displacement and stress boundaries, respectively. *P* and *p* are the external loads. *l* and *l'* are a contact point pair and n_1 and t_1 are the unit normal and tangential vectors of the contact interface, respectively. When the two objects are in contact, the contact system should satisfy the contact displacement and contact force conditions in addition to the boundary and initial conditions. This paper assumes that there is no initial gap between the two objects.

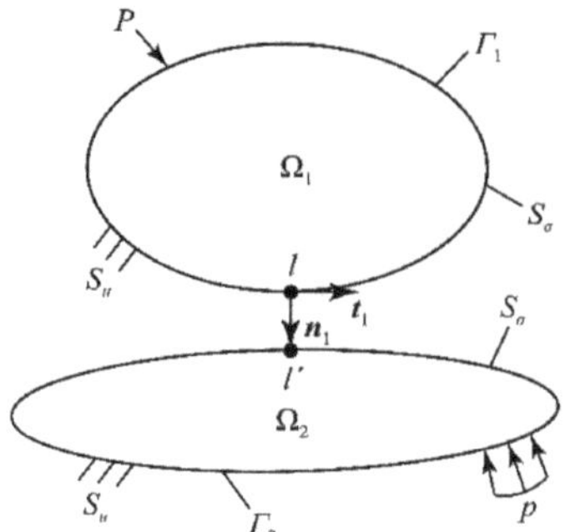

Figure 1. Contact sketch of two objects.

(1)   Contact Displacement Conditions

For two squeezed objects, it is generally considered that they will not embed into each other, so the contact point pair *l* and *l'* should satisfy the displacement condition of no embedding in the normal direction:

$$n_1^{\mathrm{T}}(U_{l'}^{t+\Delta t} - U_l^{t+\Delta t}) = 0 \tag{5}$$

where $U_{l'}^{t+\Delta t}$ and $U_l^{t+\Delta t}$ are the displacement vectors of *l'* and *l*, respectively, at time $t + \Delta t$. The direction of n_1 is from *l* to *l'*.

For a contact point pair without relative sliding, the displacement condition of no relative sliding in the tangential direction within time t–$t + \Delta t$ should be satisfied:

$$t_1^{\mathrm{T}}(U_{l'}^{t+\Delta t} - U_l^{t+\Delta t}) = t_1^{\mathrm{T}}(U_{l'}^t - U_l^t) \tag{6}$$

(2)   Contact Force Conditions

For two objects in contact, the interaction forces between the contact point pair should satisfy Newton's third law; that is, the following contact force conditions should be satisfied:

$$N_{l'}^t = -N_l^t,\ T_{l'}^t = -T_l^t \tag{7}$$

where $N_{l'}^t$ and N_l^t are the normal contact forces of *l'* and *l*, respectively. $T_{l'}^t$ and T_l^t are the tangential contact forces of *l'* and *l*, respectively.

3.2. Contact States

The above contact conditions only apply when two objects are in contact, and they do not exist at the same time. There are generally three types of contact states: bond contact, separation, and sliding contact. The static contact state is also considered in this paper.

(1)   Bond Contact State

In the bond contact state, the contact interface has good bond properties and has no relative sliding, so Equations (5)–(7) constitute the contact conditions. In addition, the normal and tangential contact forces should not exceed the ultimate tensile load and shear load, respectively:

$$\|N_l^t\| \le D^t f_c A_1 \tag{8}$$

$$\|T_l^t\| \leq D^t(\mu_\text{s}\|N_l^t\| + f_\text{c}A_1) \tag{9}$$

where f_c is the cohesive force of the contact interface, A_1 is the control area of l, and μ_s is the static friction coefficient of the contact interface.

(2)   Separation State

In the separation state, the bond properties of the contact interface are destroyed, and there is no contact between the two objects. Therefore, the contact interface is not restricted by the contact conditions, giving

$$\|N_l^t\| = \|T_l^t\| = 0 \tag{10}$$

(3)   Sliding Contact State

In the sliding contact state, the bond properties of the contact interface are destroyed, and there is relative sliding. This state is a sliding friction state, so Equations (5) and (7) constitute the contact conditions. At this point, the tangential contact force should be calculated according to the dynamic friction law:

$$\|T_l^t\| = D^t\mu_\text{d}\|N_l^t\| \tag{11}$$

where μ_d is the dynamic friction coefficient of the contact interface.

(4)   Static Contact State

In the static contact state, the bond properties of the contact interface are destroyed, but there is no relative sliding. This state is a static friction state, so Equations (5)–(7) constitute the contact conditions. At this point, the tangential contact force should not exceed the ultimate shear load:

$$\|T_l^t\| \leq D^t\mu_\text{s}\|N_l^t\| \tag{12}$$

The above contact conditions are the relationships between the contact displacement and contact force that the contact interface should satisfy when the contact state is known. In the numerical calculation process, the contact state is often not known in advance, so it is necessary to assume and judge it in each calculation step.

4. Dynamic Contact Analysis Method for Rock Mass and Fault

In the seismic loading process of a system, the contact states between the rock mass and the fault can directly affect the seismic safety of the tunnel structure. Based on the point-to-point contact type in the traditional dynamic contact force method, the point-to-surface contact type is also considered in this paper, thus improving the method for studying the nonlinear dynamic large sliding problem between a rock mass and a fault.

4.1. Fundamental Theory of the Dynamic Contact Force Method

According to the basic theory of kinematics, the dynamic balance equation for the contact nodes can be obtained after the tunnel system is discretized using finite elements:

$$M\ddot{U} + F_\text{damp} + F_\text{int} = F + R \tag{13}$$

where M is the mass matrix, U is the displacement vector, F_damp is the damping force vector, F_int is the internal force vector, which is obtained by the stress integration of the elements, F is the load vector of the seismic wave, and R is the contact force vector, $R = N + T$.

When the movement states of the contact nodes are known at time t, the integration schemes of the movement variables at time $t + \Delta t$ can be obtained, using the explicit central difference method to solve Equation (13):

$$U^{t+\Delta t} = \overline{U}^{t+\Delta t} + \Delta U^{t+\Delta t} \tag{14}$$

$$\ddot{u}^{t+\Delta t} = \frac{2(U^{t+\Delta t} - U^t)}{\Delta t} - \dot{u}^t \tag{15}$$

$$\overline{U}^{t+\Delta t} = U^t + \Delta t \dot{u}^t + \frac{\Delta t^2}{2} M^{-1}(F^t - F^t_{int} - F^t_{damp}) \tag{16}$$

$$\Delta U^{t+\Delta t} = \frac{\Delta t^2}{2} M^{-1} R^t \tag{17}$$

where $\overline{U}^{t+\Delta t}$ is the predicted displacement when ignoring the contact force R^t and $\Delta U^{t+\Delta t}$ is the modified displacement caused by R^t.

From the above integration schemes, we can see that the key to determining the movement states of the contact nodes at time $t + \Delta t$ is solving for their contact force R^t. R^t should be determined based on the contact states and the contact conditions between the rock mass and the fault within time t–$t + \Delta t$.

4.2. Solving for the Contact Force and Judging of the Contact State

The point-to-point contact is the most widely used contact type [14] and is very effective in simulating the small sliding problem, but it is difficult to simulate the large sliding problem well using this method. Based on the point-to-point contact type, the calculation formula for the contact force was derived, and methods of judging the contact state were designed. In order to simulate the large sliding problem between the rock mass and the fault, we needed to extend the point-to-point contact method to include the point-to-surface contact.

(1) Point-to-Point Contact Type

When the contact point pair has no large relative sliding, it can be considered that the contact points are of the point-to-point type (as shown in Figure 2). In order to solve the contact force R^t for this contact type, it can be assumed that the contact points are in the bond contact state. Then, they should satisfy the contact displacement conditions (Equations (5) and (6)). The normal and tangential contact gaps of the contact point pair are denoted Δ_n and Δ_t, and hence

$$\Delta_n = n_1^T(\overline{U}_{l'}^{t+\Delta t} - \overline{U}_l^{t+\Delta t}), \quad \Delta_t = t_1^T[(\overline{U}_{l'}^{t+\Delta t} - \overline{U}_l^{t+\Delta t}) - (U_{l'}^t - U_l^t)] \tag{18}$$

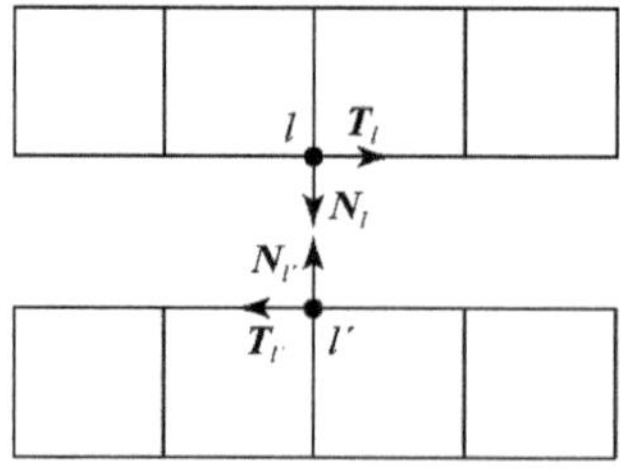

Figure 2. Point-to-point contact.

Combining Equations (5), (6), (14), (17), and (18), and according to the contact force conditions (Equation (7)), the expression for the contact force R^t_l can be derived in the form of a normal component N^t_l and a tangential component T^t_l:

$$N^t_l = \frac{2M_l M_{l'}}{(M_l + M_{l'})\Delta t^2} \Delta_n n_1, \quad T^t_l = \frac{2M_l M_{l'}}{(M_l + M_{l'})\Delta t^2} \Delta_t t_1 \tag{19}$$

where $M_{l'}$ and M_l are the lumped masses of l' and l, respectively.

If $\Delta_n > 0$, this indicates that the contact points are in a tension state. If N^t_l satisfies Equation (8), this assumption is correct; otherwise, the contact points enter a separation state, and then the contact force is modified according to Equation (10). If $\Delta_n \leq 0$, this indicates that the contact points are in a compression state. If T_l^t satisfies Equation (9), this assumption is correct; otherwise, the contact points enter a sliding contact state, and then the contact force is modified according to Equation (11).

(2) Point-to-Surface Contact Type

When the contact point pair has large relative sliding, it can be considered that the contact points are of the point-to-surface type (as shown in Figure 3), and the cohesive force is no longer considered in this case. In order to solve for the contact force R^t, it can be assumed that the contact points are in a static contact state, so they should satisfy the contact displacement conditions (Equations (5) and (6)). Assuming that l' is the projection point of l on the contact interface, the displacement of l' is given by

$$U_{l'}^t = \sum_p \Phi_p U_p^t \tag{20}$$

where p is the node number of the element to which the contact interface belongs, Φ_p is the shape-function value of p and $M_{l'}$ is the equivalent lumped mass of l', as shown below [24].

$$M_{l'} = \sum_p m_p, \quad m_p = \frac{M_p \Phi_p}{\sum_q \Phi_q^2} \tag{21}$$

where m_p is the mass contribution of p, M_p is the lumped mass of p, and q has the same meaning as p.

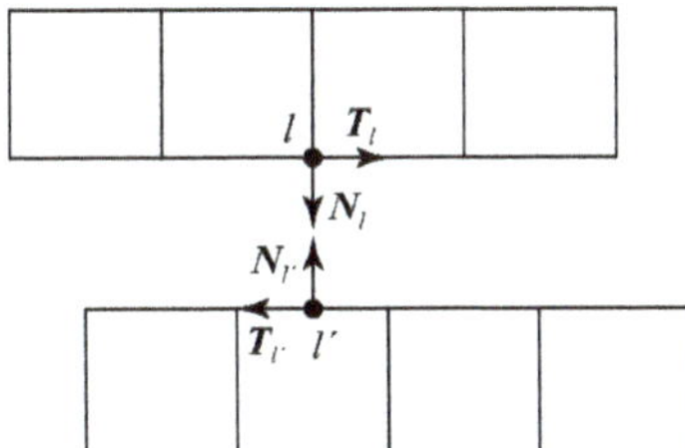

Figure 3. Point-to-surface contact.

According to the contact conditions, the expression for the contact force R_l^t can also be derived in the form of Equation (19). The contact force of p is calculated via

$$N_p^t = \frac{m_p}{M_{l'}} N_{l'}^t, \quad T_p^t = \frac{m_p}{M_{l'}} T_{l'}^t \tag{22}$$

If $\Delta_n > 0$, this indicates that the contact points enter a separation state, and then the contact force is modified according to Equation (10). If $\Delta_n \leq 0$, this indicates that the contact points are in a compression state. If T_l^t satisfies Equation (12), this assumption is correct; otherwise, the contact points enter a sliding contact state, and then the contact force is modified according to Equation (11).

4.3. Calculation Flow of the Improved Dynamic Contact Force Method

Before the seismic action is applied, the discontinuity of the finite element model can be realized by adding common nodes on the contact interface between the rock mass and the fault. The main calculation procedures for the improved dynamic contact force method are: (1) calculate the predicted displacement $\overline{U}^{t+\Delta t}$, ignoring the contact force R^t; (2) determine the contact types of all the contact point pairs; (3) use the improved method to calculate and modify R^t; (4) calculate the modified displacement $\Delta U^{t+\Delta t}$ caused by R^t;

(5) calculate the vibration deterioration coefficient $D^{t+\Delta t}$. The dynamic finite element calculation flow for the tunnel contact system at time $t + \Delta t$ is shown in Figure 4.

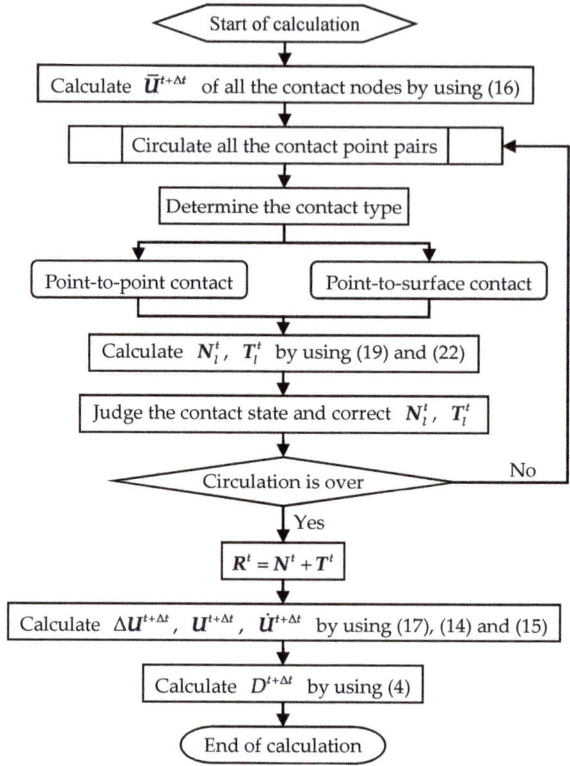

Figure 4. Calculation flow for the dynamic contact system of tunnel at time $t + \Delta t$.

5. Verification of the Method

In order to verify the accuracy of the improved dynamic contact force method for simulating large sliding displacements, the movement response of a sliding block under external loads (as shown in Figure 5) was selected for analysis [25]. The block was a cube with an edge length of 1.0 m, and the sliding surface under the block was sufficiently large. The block and the bottom surface were made of the same linear elastic material, with an elastic modulus of 0.3 GPa, a Poisson ratio of 0.3, and a density of 300 kg/m^3. The finite element mesh was composed of hexahedral elements with dimensions of 0.5 m × 0.5 m × 0.5 m. It was assumed that the gravity of the block, the cohesive force of the sliding surface, and the vibration deterioration effect of the sliding surface need not be considered. The static and dynamic friction coefficients of the sliding surface were approximately equal. In order to investigate the mutual transformation of different contact states, two calculation cases were designed, as shown in Table 1 (where t is the time).

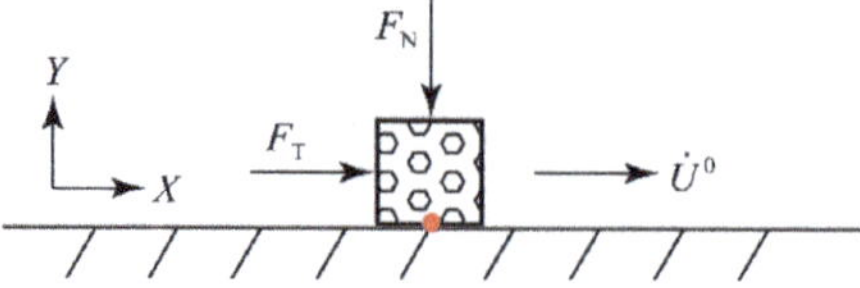

Figure 5. Contact model of the elastic block.

Table 1. Design of calculation cases.

Case	Initial Displacement U^0 (m)	Initial Velocity $\dot{U}^0$ (m·s^{-1})	Vertical Load F_N (kN)	Horizontal Load F_T (kN)	Friction Coefficient μ
1	0	0	−30	24 t	0.4
2	0	−20	−25	20	0.4

Taking the bottom center of the block as a monitoring point (as shown by the red dot in Figure 5), the velocity and displacement time histories of the block, calculated by the method developed in this paper and the analytical method, are plotted in Figures 6 and 7. In the two calculation cases, the numerical solutions of the velocity and displacement time histories of the block were basically consistent with the analytical solutions. The results show that under the above assumption, the improved dynamic contact force method is feasible, and is suitable for simulating different contact states of a contact system and the large sliding problem for contact interfaces.

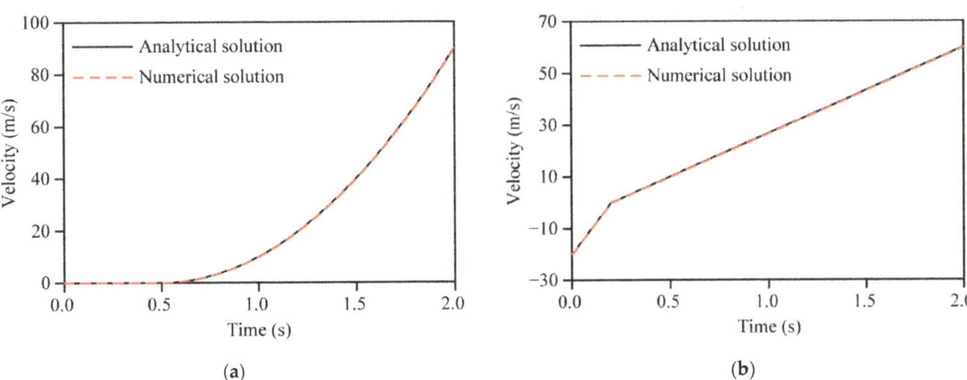

(a)   (b)

Figure 6. Velocity time histories of the elastic block: (**a**) case 1; (**b**) case 2.

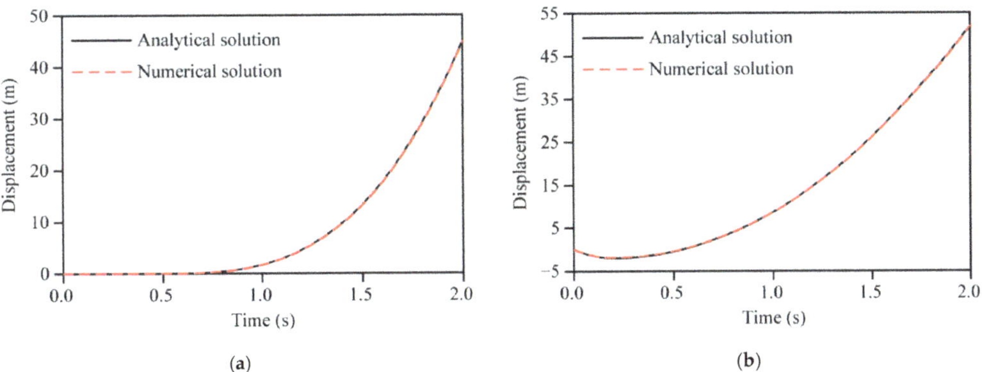

(a)   (b)

Figure 7. Displacement time histories of the elastic block: (**a**) case 1; (**b**) case 2.

6. Engineering Case Study

6.1. Engineering Profiles and Calculation Model

The water diversion project of Central Yunnan is a super-large inter-basin water transfer project, which takes water from the Shigu segment of the upper reaches of the Jinsha River to solve the water shortage problem in the Central Yunnan areas. The control project for the main canal is the Xianglushan tunnel in Dali section I, with a total length of 63.43 km and a buried depth of more than 300 m. Many active fault zones exist along the line, most of which intersect with the line at medium or large angles. The basic earthquake intensity degree of the tunnel area is degree 8, and the peak acceleration of the horizontal ground motion with 10% exceedance probability in 50 years is 0.2–0.3 g. A tunnel section with a fault and a buried depth of 400 m was selected for the seismic calculation in this study. The rock mass is type IV, and the tunnel diameter is 9.8 m. The thickness of the lining support is 55 cm, with a C30 concrete structure.

Due to the great depth of the tunnel, the number of elements would be very large and the dynamic calculation would take a great deal of time if the 3D finite element model was built to the ground surface. Therefore, a depth of only 55 m was used at the top of the tunnel, to save calculation time. A normal fault with a thickness of 20 m and a dip angle of 60° was considered, with an inclination parallel to the tunnel axis. The finite element model was composed of 69,888 rock mass elements, 17,472 fault elements, and 8640 lining elements, all of which were eight-node hexahedrons, as shown in Figure 8. The maximum mesh size was within 5 m, which satisfies the accuracy requirement of the dynamic calculation. The initial contact node pairs were set between the hanging wall, the fault, and the footwall, respectively. It should be noted that the contact interface between the concrete lining and the fault or the rock mass was not considered.

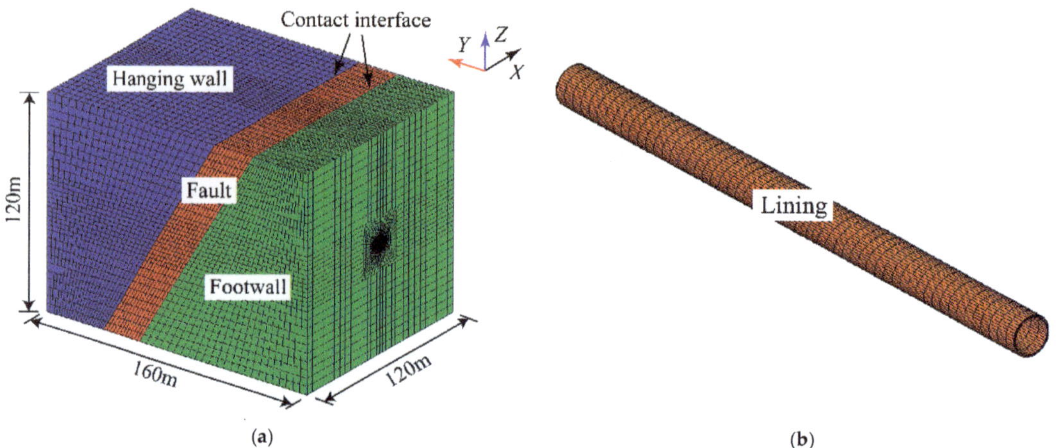

Figure 8. Three-dimensional calculation model of the tunnel with a normal fault: (**a**) whole model; (**b**) lining.

Based on a comprehensive analysis of in situ stress test results in the tunnel area, the lateral pressure coefficients of the rock mass were taken as: $k_X = 1.2$, $k_Y = 0.74$, $k_Z = 1.0$. The values of the mechanical parameters of the tunnel materials are provided in Table 2. The cohesive force of the contact interface between the rock mass and the fault was taken as 0.35 MPa, and the static and dynamic friction coefficients were both taken as 0.53. The undetermined coefficients in Equation (4) were taken as: $\alpha = 0.883$, $\beta = 0.02$, $\lambda = 0.032$, $P_0 = 0.8$, $\xi = 5$. As the research was limited by current test conditions, the values of these undetermined coefficients were mainly based on the test fitting curves in [19,23].

Table 2. Mechanical parameters of tunnel materials.

Material	Elastic Modulus (GPa)	Poisson Ratio	Internal Friction Angle (°)	Cohesive Force (MPa)	Tensile Strength (MPa)	Compressive Strength (MPa)	Density (g·cm^{-3})
Fault	0.3	0.33	24.2	0.1	0.50	10.0	2.0
Lining	30.0	0.17	46.0	2.0	1.43	12.7	2.5
Rock	5.0	0.29	35.0	0.6	1.30	40.0	2.7

6.2. Calculation Conditions

Based on the improved dynamic contact force method, a seismic response process simulation of the Xianglushan tunnel with a normal fault was performed, using the dynamic time history finite element analysis program for an underground cavern [26]. The same elastoplastic damage constitutive relation was applied to the rock mass and the concrete lining, and the Mohr–Coulomb yield function with a tensile cut-off limit was applied as the yield condition, expressed as

$$f^{\mathrm{s}} = \sigma_1 - \sigma_3 N_\varphi + 2(1-D)c\sqrt{N_\varphi}, f^{\mathrm{t}} = \sigma_3 - \sigma^{\mathrm{t}} \tag{23}$$

where $f^{\mathrm{s}} = 0$ and $f^{\mathrm{t}} = 0$ represent the shear and tensile yield conditions, respectively, and σ_3 and σ_1 are the maximum and minimum principal stresses, respectively. It was assumed that tensile stress is positive and compressive stress is negative. In addition, c is the cohesive force, σ^{t} is the tensile strength and N_φ is a parameter expressed as

$$N_\varphi = \frac{1 + \sin\varphi}{1 - \sin\varphi} \tag{24}$$

where φ is the internal friction angle. D is the damage coefficient, which is expressed as

$$D = \sqrt{D_1^2 + D_2^2 + D_3^2} \tag{25}$$

$$D_i = 1 - \exp(-R\sqrt{\varepsilon_i^p \varepsilon_i^p}), \; i = 1, 2, 3 \tag{26}$$

where D_i is the damage coefficient of the i-th principal stress direction, ε_i^p is the cumulative plastic deviator strain of the i-th principal strain direction and R is a dimensionless constant.

A free field artificial boundary condition was applied at the four vertical boundaries, and a viscoelastic artificial boundary condition was applied at the top and bottom of the finite element model.

In this study, a representative Kobe NS wave was selected for input to the model in the form of acceleration time histories. A duration section of 15 s with high amplitude was adopted to shorten the calculation time. Then, the amplitude of the input wave was adjusted to 0.15 g to meet the requirements of seismic fortification intensity and the amplitude reduction of a deep tunnel. The processed acceleration time histories are shown in Figure 9, where only the x-direction seismic excitation is considered.

In addition, a contrasting calculation case was designed, with no contact node pairs between the rock mass and the fault. In other words, the rock mass and the fault were regarded as a continuous structure. Then, the seismic responses of the tunnel in the two calculation cases (considering or ignoring the dynamic interaction between the rock mass and the fault (with RFI or without RFI)) were analyzed comparatively. In order to monitor the displacement and stress time histories of the lining, three monitoring points were set at typical parts of the lining: the bottom arch, the haunch, and the top arch at the middle position of the fault. It should be noted that the tunnel excavation and the lining support calculations should be performed before the seismic action is applied.

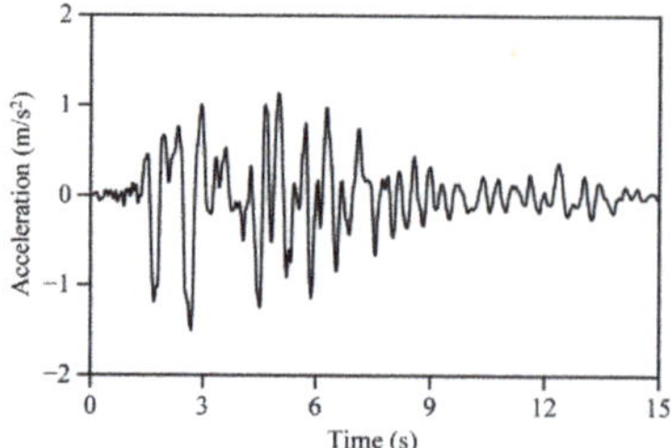

Figure 9. Time history of x-direction acceleration of the input wave.

6.3. Analysis of Calculation Results

6.3.1. Relative Movement and Seismic Deterioration Analysis of the Contact Interface

In the seismic loading process of the system, the relative displacement and relative velocity between the rock mass and the fault leads to the seismic deterioration of the contact interface. A contact node pair close to the lining haunch was selected at the contact interface between the hanging wall and the fault, to monitor the relative movement characteristics of the rock mass and the fault. With the dynamic interaction included, the time histories for the relative x-direction movement of the contact interface between the hanging wall and the fault are plotted in Figure 10, and the time history of the vibration deterioration coefficient of the contact interface is plotted in Figure 11.

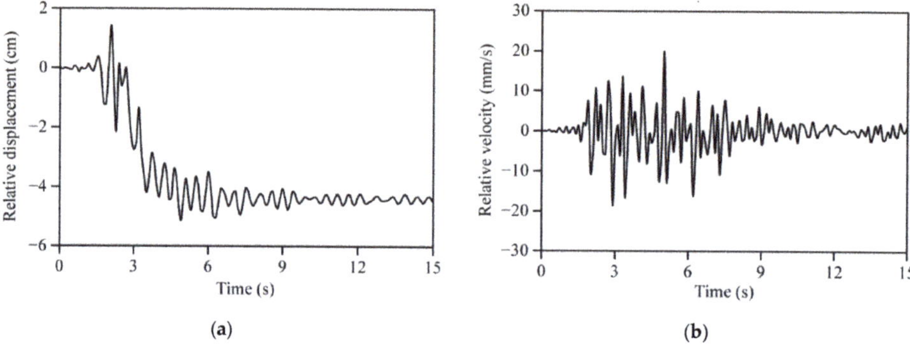

(a)   **(b)**

Figure 10. Time histories of the relative movement of the contact interface between the hanging wall and the fault: (a) displacement; (b) velocity.

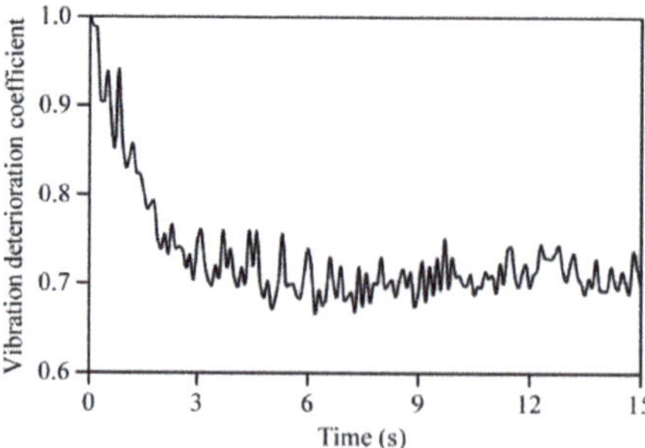

Figure 11. Time history of the vibration deterioration coefficient of the contact interface.

We can see from Figure 10a that the relative displacement of the contact interface was very small at time 0–1.5 s, and it fluctuated violently and increased sharply at time 1.5–6.0 s.

The maximum relative displacement was −5.14 cm, appearing at time 4.9 s. After time 6.0 s, the relative displacement basically remained at −4.40 cm. This indicates that the movements of the rock mass and the fault were not synchronous under a large seismic action, and an obvious dislocation displacement appeared between them. That is to say, the rock mass and the fault were in a sliding contact state after time 6.0 s. We can see from Figure 10b that the relative velocity of the contact interface presented similar fluctuating change laws, due to the fluctuation of the input wave: the relative velocity fluctuated violently at time 1.5–6.0 s and fluctuated gently after time 6.0 s.

We can see from Figure 11 that the vibration deterioration coefficient of the contact interface presented clear negative exponential attenuation laws over time: the coefficient decreased rapidly at time 0–1.5 s, then decreased to a certain extent at time 1.5–6.0 s, tending to be stable after time 6.0 s. This indicates that the seismic deterioration effect of the contact interface accumulated gradually over time, and its deterioration degree can be directly reflected in the value of the vibration deterioration coefficient.

6.3.2. Displacement Analysis of the Lining

With the dynamic interaction between the rock mass and the fault considered or ignored, the time histories of the x-direction displacement of the lining monitoring points in the two calculation cases are plotted in Figure 12.

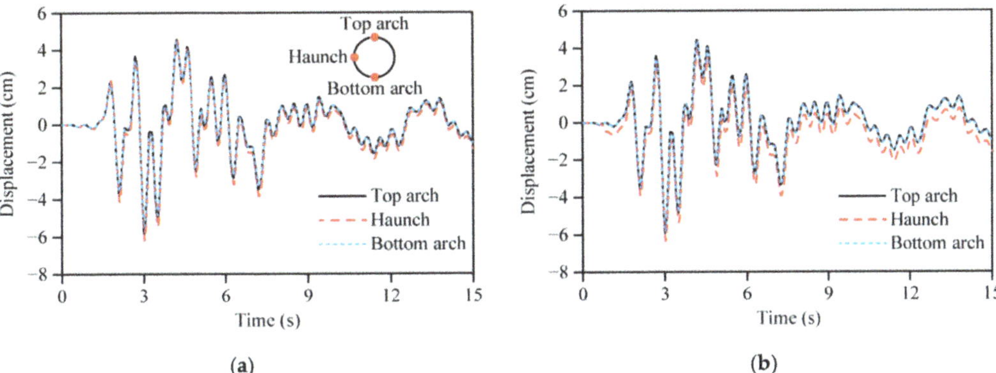

Figure 12. Time histories of the x-direction displacement of the monitoring points: (**a**) without RFI; (**b**) with RFI.

We can see from Figure 12 that the movement laws of the three typical parts were basically the same, and only the magnitudes at each point in time were somewhat different. On the whole, the displacement time histories of the bottom arch and the top arch were basically the same, while the displacement of the haunch along the x-direction was larger than that of the other parts. The main reason for this is that, influenced by the spatial structure of the tunnel, the seismic response of the haunch was more intense than that of the other parts under a transverse seismic wave. When the dynamic interaction between the rock mass and the fault was ignored, the maximum displacement of the haunch was −6.16 cm, appearing at time 3.0 s, and the maximum relative displacement between the haunch and the bottom arch was −0.33 cm. When the dynamic interaction was considered, the maximum displacement of the haunch was −6.31 cm, also appearing at time 3.0 s, and the maximum relative displacement between the haunch and the bottom arch was −0.64 cm. In the two calculation cases, the displacement responses of the lining are clearly different. The main reason for this is that an obvious transverse dislocation displacement appears between the rock mass and the fault under seismic action, which has an important influence on the seismic response of the lining monitoring section, especially the displacement of the haunch. This indicates that the transverse dislocation displacement between the rock mass and the fault leads to large relative deformations between different parts of the lining.

Therefore, considering the dynamic interaction between the rock mass and the fault can more objectively reflect the displacement response characteristics of the tunnel structure.

According to the above analysis, the displacement of the lining haunch along the x-direction reached a maximum at time 3.0 s. With the dynamic interaction between the rock mass and the fault considered, the changing laws of the transverse maximum displacement of the haunch along the tunnel axis at time 3.0 s are plotted in Figure 13. Here, the minus sign indicates the direction.

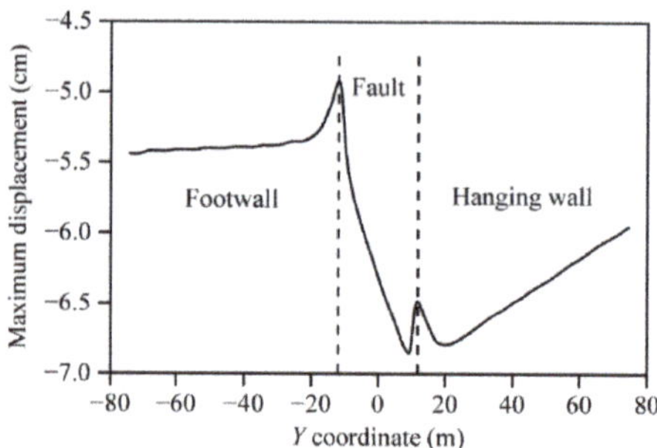

Figure 13. Changing laws of the maximum displacement of the haunch along the tunnel axis.

We can see from Figure 13 that the maximum displacements of the haunch at different y coordinates of the footwall were basically the same. From the footwall to the fault, the displacement decreased sharply at the contact interface, and increased rapidly at the fault. From the fault to the hanging wall, the displacement appeared to change suddenly at the contact interface, decreasing slowly at the hanging wall. This indicates that the displacement of the lining was greatly affected by the fault dislocation, especially near the contact interface. Compared with the lining at the footwall, the fault was more likely to affect the lining at the hanging wall, which was characterized by a larger displacement.

6.3.3. Stress and Damage Analysis of the Lining

In view of the fact that the tensile strength of concrete is much smaller than its compressive strength, only the changing laws of the tensile stress of the lining were analyzed in this study, which is reflected in the maximum principal stress. With the dynamic interaction between the rock mass and the fault considered or ignored, the time histories of the maximum principal stress of the lining monitoring points in the two calculation cases are plotted in Figure 14.

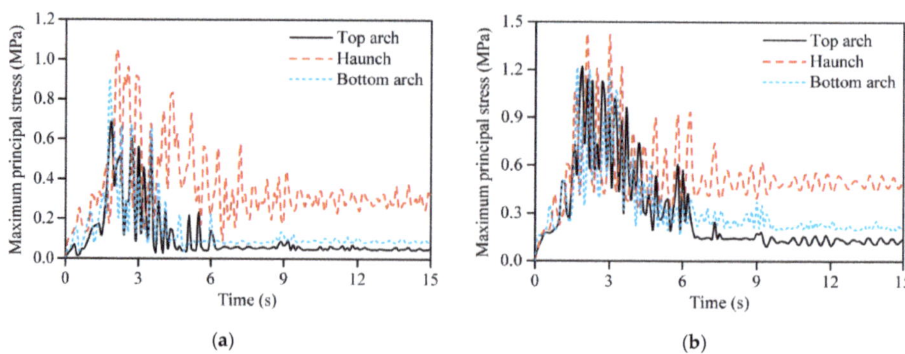

Figure 14. Time histories of the maximum principal stress of the monitoring points: (**a**) without RFI; (**b**) with RFI.

We can see from Figure 14 that the changing laws of the maximum principal stress of the three typical parts were basically the same: the stress increased slowly at time 0–1.5 s, then changed and increased rapidly at time 1.5–6.0 s, changing little after time 6.0 s. On the whole, the tensile stress of the haunch was larger than that of the other parts. The maximum tensile stress of the lining was 1.05 MPa when the dynamic interaction between the rock mass and the fault was ignored, while the value was 1.43 MPa when the dynamic interaction was considered, which is clearly larger than that in the former case. In addition, when the dynamic interaction was considered, the maximum tensile stress of the haunch reached the tensile strength of concrete, and the haunch may then suffer damage. This indicates that the stress of the lining is greatly affected by the dislocation displacement between the rock mass and the fault.

According to the above analysis, tensile damage appears on the lining when the dynamic interaction between the rock mass and the fault is considered. Based on the damage constitutive relation of concrete, the distribution of the damage coefficient of the lining structure after the earthquake can be obtained, as shown in Figure 15, and the changing laws of the damage coefficient of the lining haunch along the tunnel axis after the earthquake are plotted in Figure 16.

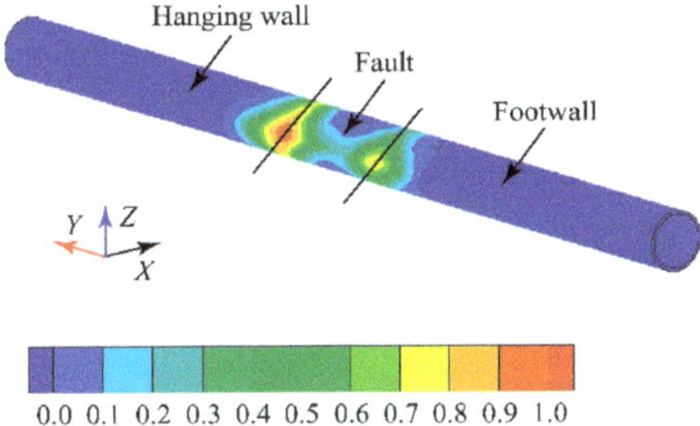

Figure 15. Distribution of the damage coefficient of the lining after the earthquake.

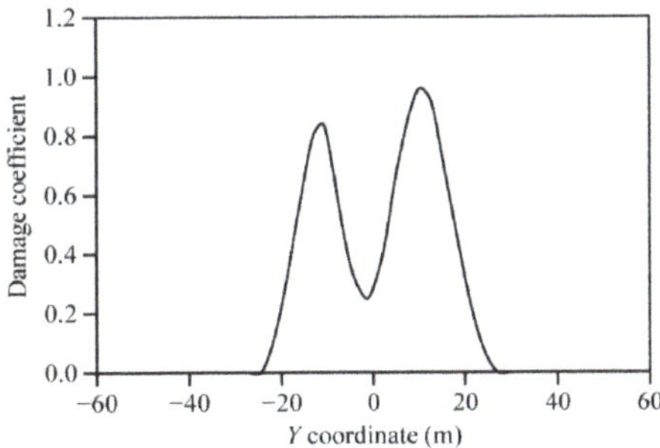

Figure 16. Changing laws of the damage coefficient of the lining haunch along the tunnel axis.

We can see from Figures 15 and 16 that the damage area of the lining was mainly distributed in a certain range near both sides of the fault and at the parts where the fault passes through, especially at the haunch. From the footwall to the fault, the damage coefficient of the haunch first increased and then decreased. From the fault to the hanging wall, the damage coefficient of the haunch also first increased and then decreased. Compared with the other parts, the damage to the lining at the contact interface between the rock mass and the fault was the most serious. Compared with the contact interface between the footwall and the fault, the damage to the lining at the contact interface between the hanging wall and the fault was more serious. The local damage coefficient of the haunch at this part was close to 1.0, and the haunch may then crack. Based on the above analysis, the influences of fault dislocations should be considered in the seismic design of a tunnel structure through a fault, from the perspective of engineering safety. In addition, according to the distribution length of the damage area of the lining, the longitudinal layout range of seismic fortification for the tunnel structure can be determined.

6.4. Discussion

The seismic response of the tunnel through fault is affected by many factors, including fault thickness and fault dip angle. In this section, when the influences of different fault thicknesses on the seismic response of the tunnel were considered, the fault dip angle was uniformly set at 60°. When the influences of different fault dip angles on the seismic response of the tunnel were considered, the fault thickness was uniformly set at 20 m. In all the above calculation cases, the other calculation conditions remained consistent, including the fact that the dynamic interaction between the rock mass and the fault was considered.

6.4.1. Influences of Fault Thickness on Seismic Response of the Lining

The maximum displacements and maximum tensile stresses of the lining monitoring points under different fault thicknesses are plotted in Figure 17, and the damage coefficients of the haunch monitoring point under different fault thicknesses are plotted in Figure 18.

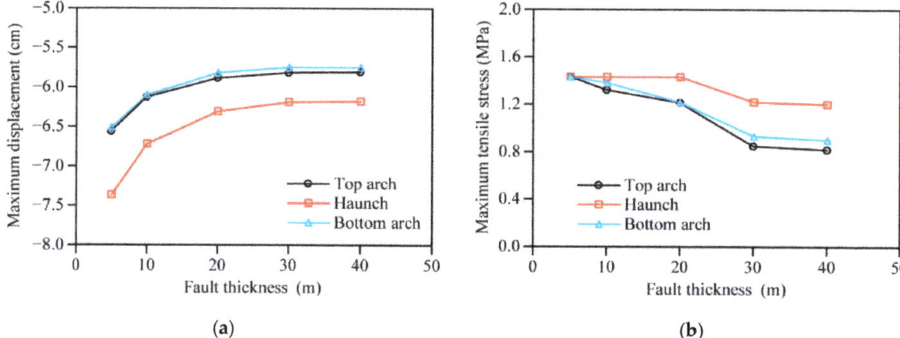

(a)

(b)

Figure 17. Maximum response values of the monitoring points under different fault thicknesses: (a) displacement; (b) tensile stress.

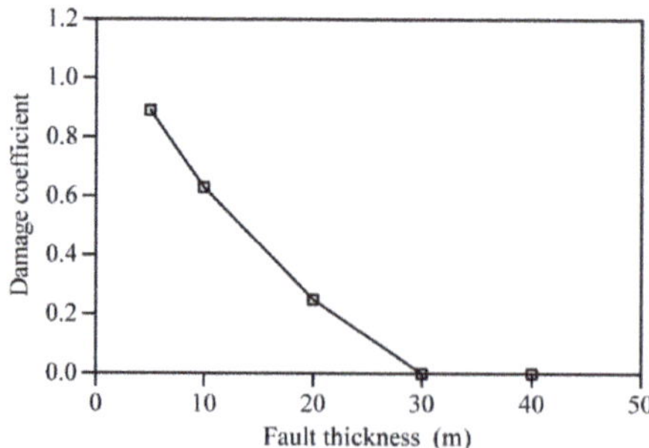

Figure 18. Damage coefficients of the haunch monitoring point under different fault thicknesses.

We can see from Figure 17 that as the fault thickness increased, the maximum displacements of the three typical parts decreased gradually, tending to be stable when the thickness exceeded 30 m. Similarly, the changing of the maximum tensile stresses with fault thickness showed the same laws. This is mainly because the larger the fault thickness, the farther the lining monitoring section from the contact interface between the rock mass and the fault, so that the influence of the fault dislocation on the monitoring points is smaller.

When the fault thickness exceeds a certain size, the influence of the fault dislocation on the monitoring points can be ignored. It should be noted that the maximum tensile stress of the haunch reached the tensile strength of concrete when the fault thickness did not exceed 20 m, and this part may then suffer damage. Furthermore, we can see from Figure 18 that as the fault thickness increased, the damage coefficient of the lining haunch decreased gradually, remaining at zero when the thickness exceeded 30 m.

6.4.2. Influences of Fault Dip Angle on Seismic Response of the Lining

The maximum displacements and maximum tensile stresses of the lining monitoring points under different fault dip angles are plotted in Figure 19, and the damage coefficients of the haunch monitoring point under different fault dip angles are plotted in Figure 20.

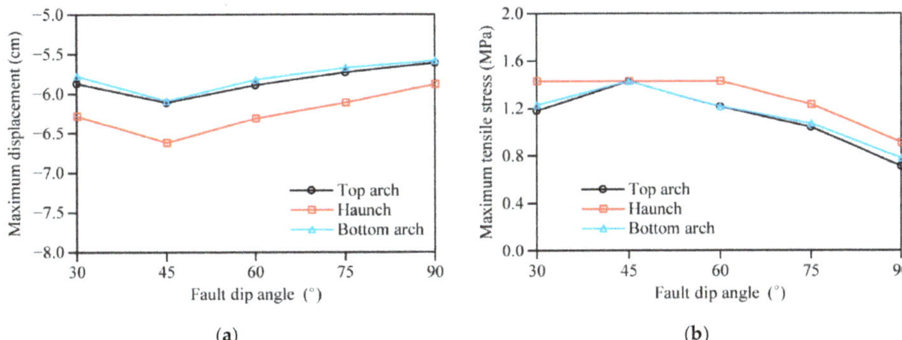

Figure 19. Maximum response values of the monitoring points under different fault dip angles: (**a**) displacement; (**b**) tensile stress.

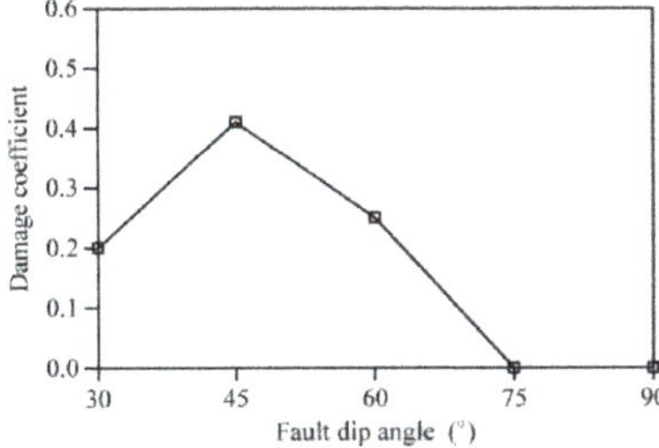

Figure 20. Damage coefficients of the haunch monitoring point under different fault dip angles.

We can see from Figure 19 that as the fault dip angle increased, both the maximum displacements and maximum tensile stresses of the three typical parts first increased and then decreased, reaching a maximum at 45°. This indicates that the fault dip angle had a great influence on the seismic response of the lining monitoring section. It should be noted that the maximum tensile stress of the haunch reached the tensile strength of concrete when the fault dip angle did not exceed 60°, and this part may then suffer damage. Furthermore, we can see from Figure 20 that as the fault dip angle increased, the damage coefficient of the lining haunch first increased and then decreased, reaching a maximum at 45°. When the fault dip angle exceeded 75°, the damage coefficient of the lining haunch remained zero.

7. Conclusions

(1) Considering the dynamic deterioration characteristics of a rock structural plane under seismic action, a mathematical model reflecting the seismic deterioration effect of the contact interface between the rock mass and the fault was established. Based on the point-to-point contact type in the traditional dynamic contact force method, the point-to-

surface contact type was also considered, and an improved dynamic contact force method considering the large sliding characteristics of the contact interface was established. Then, a nonlinear dynamic simulation method for the rock–fault contact system was developed. The calculation flow of the improved dynamic contact force method was designed, and the accuracy of the method was verified using the example of a sliding elastic block.

(2) Based on the improved dynamic contact force method, the nonlinear seismic damage characteristics of a deep tunnel through a normal fault were studied. The relative movement and seismic deterioration effect of the contact interface between the rock mass and the fault, and the characteristics of displacement, stress, and damage to the lining, were analyzed. The results indicated that the movements of the rock mass and the fault were not synchronous under large seismic action, and the seismic deterioration effect of the contact interface accumulated gradually over time. The proposed method can effectively simulate the vibration deterioration degree of the contact interface and the nonlinear large sliding problem of the rock–fault contact system.

(3) When the dynamic interaction between the rock mass and the fault was considered, the displacement and stress responses of the lining were greatly affected by the transverse dislocation displacement between the two, shown by a clear increase in the values. Compared with the lining at the footwall, the displacement of the lining at the hanging wall was more easily affected by the fault. Tensile damage appeared on the lining in the seismic loading process, and the damage area of the lining was mainly distributed in a certain range near both sides of the fault and at the parts where the fault passes through, especially at the haunch. Compared with the contact interface between the footwall and the fault, the damage to the lining at the contact interface between the hanging wall and the fault was more serious. This indicates that considering the dynamic interaction between rock mass and fault can more objectively reflect the seismic response characteristics of the tunnel structure.

(4) The influences of different fault thicknesses and different fault dip angles on the seismic response of the tunnel structure were discussed. As the fault thickness increased, both the maximum displacements and the maximum tensile stresses of the lining monitoring points decreased gradually, tending to be stable when the thickness exceeded 30 m. The damage coefficient of the lining haunch decreased gradually, and remained zero when the fault thickness exceeded 30 m. As the fault dip angle increased, both the maximum displacements and maximum tensile stresses of the lining monitoring points first increased and then decreased, reaching a maximum at 45°. The damage coefficient of the lining haunch first increased and then decreased, reaching a maximum when the fault dip angle was 45° and remaining zero when the fault dip angle exceeded 75°.

Author Contributions: Conceptualization, G.L.; methodology, Y.Z.; software, M.X.; validation, Y.Z.; formal analysis, G.L.; investigation, J.R.; resources, M.X.; data curation, J.R.; writing—original draft preparation, G.L.; writing—review and editing, G.L.; visualization, J.R.; supervision, Y.Z. and M.X.; project administration, G.L.; funding acquisition, Y.Z. and M.X. All authors have read and agreed to the published version of the manuscript.

Funding: This research was supported by the National Key Research and Development Program of China (grant number 2017YFC0404901), and the National Natural Science Foundation of China (grant number 52079097).

Institutional Review Board Statement: Not applicable.

Informed Consent Statement: Not applicable.

Data Availability Statement: Not applicable.

Conflicts of Interest: The authors declare no conflict of interest.

References

1. Huang, J.Q.; Zhao, M.; Du, X.L. Non-linear seismic responses of tunnels within normal fault ground under obliquely incident P waves. *Tunn. Undergr. Space Technol.* **2017**, *61*, 26–39. [CrossRef]
2. Wang, Z.Z.; Zhang, Z. Seismic damage classification and risk assessment of mountain tunnels with a validation for the 2008 Wenchuan earthquake. *Soil Dyn. Earthq. Eng.* **2013**, *45*, 45–55. [CrossRef]
3. Yan, G.M.; Gao, B.; Shen, Y.S.; Zheng, Q.; Fan, K.X.; Huang, H.F. Shaking table test on seismic performances of newly designed joints for mountain tunnels crossing faults. *Adv. Struct. Eng.* **2020**, *23*, 248–262. [CrossRef]
4. Shen, Y.S.; Gao, B.; Yang, X.M.; Tao, S.J. Seismic damage mechanism and dynamic deformation characteristic analysis of mountain tunnel after Wenchuan earthquake. *Eng. Geol.* **2014**, *180*, 85–98. [CrossRef]
5. Li, T.B. Damage to mountain tunnels related to the Wenchuan earthquake and some suggestions for aseismic tunnel construction. *Bull. Eng. Geol. Environ.* **2012**, *71*, 297–308. [CrossRef]
6. Yu, H.T.; Chen, J.T.; Bobet, A.; Yuan, Y. Damage observation and assessment of the Longxi tunnel during the Wenchuan earthquake. *Tunn. Undergr. Space Technol.* **2016**, *54*, 102–116. [CrossRef]
7. Fang, L.; Jiang, S.P.; Lin, Z.; Wang, F.Q. Shaking table model test study of tunnel through fault. *Rock Soil Mech.* **2011**, *32*, 2709–2713.
8. Liu, Y.; Gao, F. Experimental study on the dynamic characteristics of a tunnel-crossing fault using a shake-table test. *J. Vib. Shock.* **2016**, *35*, 160–165.
9. Fan, L.; Chen, J.L.; Peng, S.Q.; Qi, B.X.; Zhou, Q.W.; Wang, F. Seismic response of tunnel under normal fault slips by shaking table test technique. *J. Cent. South. Univ.* **2020**, *27*, 1306–1319. [CrossRef]
10. Li, L.; He, C.; Geng, P.; Cao, D.J. Analysis of seismic dynamic responses of tunnel through fault zone in high earthquake intensity area. *J. Chongqing Univ. Nat. Sci. Ed.* **2012**, *35*, 92–98.
11. Yang, Z.H.; Lan, H.X.; Zhang, Y.S.; Gao, X.; Li, L.P. Nonlinear dynamic failure process of tunnel-fault system in response to strong seismic event. *J. Asian Earth Sci.* **2013**, *64*, 125–135. [CrossRef]
12. Ardeshiri-Lajimi, S.; Yazdani, M.; Langroudi, A.A. Control of fault lay-out on seismic design of large underground caverns. *Tunn. Undergr. Space Technol.* **2015**, *50*, 305–316. [CrossRef]
13. Liu, J.B.; Sharan, S.K. Analysis of dynamic contact of cracks in viscoelastic media. *Comput. Methods Appl. Mech. Eng.* **1995**, *121*, 187–200. [CrossRef]
14. Yang, Y.; Chen, J.T.; Xiao, M. Analysis of seismic damage of underground powerhouse structure of hydropower plants based on dynamic contact force method. *Shock. Vib.* **2014**, *2014*, 1–13. [CrossRef]
15. Zhou, H.; Xiao, M.; Yang, Y.; Liu, G.Q. Seismic response analysis method for lining structure in underground cavern of hydropower station. *KSCE J. Civ. Eng.* **2019**, *23*, 1236–1247. [CrossRef]
16. Hutson, R.W.; Dowding, C.H. Joint asperity degradation during cyclic shear. *Int. J. Rock Mech. Min. Sci.* **1990**, *27*, 109–119. [CrossRef]
17. Hou, D.; Rong, G.; Yang, J.; Zhou, C.B.; Peng, J.; Wang, X.J. A new shear strength criterion of rock joints based on cyclic shear experiment. *Eur. J. Environ. Civ. Eng.* **2016**, *20*, 180–198. [CrossRef]
18. Yang, Z.Y.; Di, C.C.; Yen, K.C. The effect of asperity order on the roughness of rock joints. *Int. J. Rock Mech. Min. Sci.* **2001**, *38*, 745–752. [CrossRef]
19. Ni, W.D.; Tang, H.M.; Liu, X.; Wu, Y.P. Dynamic stability analysis of rock slope considering vibration deterioration of structural planes under seismic loading. *Chin. J. Rock Mech. Eng.* **2013**, *32*, 492–500.
20. Crawford, A.M.; Curran, J.H. The influence of shear velocity on the frictional resistance of rock discontinuities. *Int. J. Rock Mech. Min. Sci.* **1981**, *18*, 505–515. [CrossRef]
21. Wang, Z.; Gu, L.L.; Shen, M.R.; Zhang, F.; Zhang, G.K.; Deng, S.X. Influence of shear rate on the shear strength of discontinuities with different joint roughness coefficients. *Geotech. Test. J.* **2020**, *43*, 683–700. [CrossRef]
22. Lee, H.S.; Park, Y.J.; Cho, T.F.; You, K.H. Influence of asperity degradation on the mechanical behavior of rough rock joints under cyclic shear loading. *Int. J. Rock Mech. Min. Sci.* **2001**, *38*, 967–980. [CrossRef]
23. Li, H.B.; Feng, H.P.; Liu, B. Study on strength behaviors of rock joints under different shearing deformation velocities. *Chin. J. Rock Mech. Eng.* **2006**, *25*, 2435–2440.
24. Wang, F.J.; Cheng, J.G.; Yao, Z.H.; Huang, C.J.; Kou, Z.J. A new contact algorithm for numerical simulation of structure crashworthiness. *Eng. Mech.* **2002**, *19*, 130–134.
25. Zhang, L.H.; Liu, T.Y.; Li, Q.B.; Chen, T. Modified dynamic contact force method under reciprocating load. *Eng. Mech.* **2014**, *31*, 8–14.
26. Zhang, Z.G.; Xiao, M.; Chen, J.T. Simulation of earthquake disaster process of large-scale underground caverns using three-dimensional dynamic finite element method. *Chin. J. Rock Mech. Eng.* **2011**, *30*, 509–523.

Article

Evaluation of Ground Motion Amplification Effects in Slope Topography Induced by the Arbitrary Directions of Seismic Waves

Chao Yin [1,*], Wei-Hua Li [2] and Wei Wang [3]

[1] State Key Laboratory of Mechanical Behavior and System Safety of Traffic Engineering Structures, Shijiazhuang Tiedao University, Shijiazhuang 050043, China
[2] School of Civil Engineering, Beijing Jiaotong University, Beijing 100044, China; whli@bjtu.edu.cn
[3] Key Laboratory of Roads and Railway Engineering Safety Control of Ministry of Education, Shijiazhuang Tiedao University, Shijiazhuang 050043, China; wangweiuuu@163.com
*   Correspondence: robinyc@stdu.edu.cn; Tel.: +86-311-8793-9232

Abstract: The incident directions of seismic waves can change the ground motions of slope topography. To elaborate on the influences of the directions of seismic waves, a dynamic analysis of the slope topography was performed. Seismic waves were input using an equivalent nodal force method combined with a viscous-spring artificial boundary. The amplification of ground motions in double-faced slope topographies was discussed by varying the angles of incidence. Meanwhile, the components of seismic waves (P waves and SV waves), slope materials and slope geometries were all investigated with various incident earthquake waves. The results indicated that the pattern of the amplification of SV waves was stronger than that of P waves in the slope topography, especially in the greater incident angels of the incident waves. Soft materials intensely aggravate the acceleration amplification, and more scattered waves are produced under oblique incident earthquake waves. The variations in the acceleration amplification ratios on the slope crest were much more complicated at oblique incident waves, and the ground motions were underestimated by considering only the vertical incident waves. Therefore, in the evaluation of ground motion amplification of the slope topography, it is extremely important to consider the direction of incident waves.

Keywords: directions of incident waves; slope topography; amplification effects; ground motions; equivalent nodal forces

Citation: Yin, C.; Li, W.-H.; Wang, W. Evaluation of Ground Motion Amplification Effects in Slope Topography Induced by the Arbitrary Directions of Seismic Waves. *Energies* **2021**, *14*, 6744. https://doi.org/10.3390/en14206744

Academic Editor: Manoj Khandelwal

Received: 20 August 2021
Accepted: 13 October 2021
Published: 16 October 2021

Publisher's Note: MDPI stays neutral with regard to jurisdictional claims in published maps and institutional affiliations.

1. Introduction

Earthquake-induced landslides are among the most hazardous secondary effects of strong seismicity in mountainous regions. Several historical cases highlight the potentially devastating consequences in the last 30 years [1–6] because nearly one-third of deaths are caused by earthquake-induced landslides among all disasters [7–10]. Many researchers have indicated that topographic effects have possible interactions between landslide mechanisms and triggering conditions [11–16]. Thus, the topographic effects, represented by the ground motions on the slope model, should be better understood.

Generally, convex topographies such as mountains, slopes and individual ridges lead to intense aggravation of the seismic responses irregularly along the ground surface [17]. Many acceleration-time histories recorded by the seismometers near the epicenter all proved that the peak acceleration at the slope crest showed intense amplification based on instrumental data from strong earthquakes such as the 1987 California earthquake, the 1994 Northridge earthquake, the 1995 Egion earthquake and the 2008 Wenchuan earthquake [12,18–20]. In addition, statistical analysis of the seismological stations in Iran and Israel showed that the acceleration amplification ratios with respect to the slope crest and ground could reach as much as four times [21,22]. These measured records confirmed that the slope topography modified the seismic ground motions.

In the analysis of topographic effects, both analytical and numerical methods were utilized [23–34]. Nevertheless, the analytical method (e.g., Newmark's method) can neither quantify the plastic strains of the non-shearing models nor consider the interaction between the models and the seismic waves in terms of ground motion amplification or de-amplification owing to irregular topographic patterns [17]. Conversely, the complicated stress-strain relationships can be described clearly based on the numerical method, and the expected plastic strain effects of the model can be quantified owing to seismic events [35]. Thus, for convex topographies (mountains, slopes and individual ridges), the most common method is a numerical simulation, such as finite difference methods [13,36–39], finite element methods [40–45], boundary element methods [37,46,47], generalized consistent transmitting boundary methods [29,30] and distinct element methods [48].

Many researchers have analyzed the ground motions in slope topography [13,36,37,40–45,49,50]; however, they preferred the vertical directions of incident waves rather than considering the influences of the incident angles. According to the regression analysis of numerous seismological records in America and Japan, the incident angles near the ground are usually oblique [51,52]. Undoubtedly, the oblique incident directions change the seismic wave paths in the slope model, and the different seismic responses lead to the irregularly intense amplification of the ground motions and aggravation of the slope instability [11,30]. Thus, the varied angles of incident waves definitely influence the amplification of the slope topography, and overlooking the wave inclination oversimplifies the analyses of the seismic responses on the ground motions. Even though this problem has been discussed by some researchers [30,35], there has been little systematic analysis of the influences of the incident directions on the ground motions in slope topography, such as the impacts of slope materials and sizes under oblique angles of incident waves.

Based on the analyses above, this study aimed to investigate the impact of the incident directions of seismic waves on the ground motions in slope topography. The remainder of this paper is organized as follows. The problem is proposed and the slope model is established in Section 2. The derivation of the input method in the numerical simulation with arbitrary directions of incident waves is presented in Section 3. In Section 4, the influencing factors (wave patterns, slope materials and slope geometries) are discussed separately under a double-faced slope topography with varying angles of incident waves. Finally, the conclusions and future prospects are presented in Section 5.

2. Problem Layout and Model Establishment

Topography has an important influence on the amplification of ground motions. The convex topography, which includes slopes, individual hills and mountains, leads to irregular amplification of the topographic effects. In this study, the configurations of the slopes, hills and mountains were simplified as double-faced slope topographies, and the universality of the inclinations of seismic waves that enter the site was considered. The impacts of the topographic effects were analyzed based on the double-faced slope topographies and the site effects were investigated under arbitrary directions of incident waves with other factors (e.g., components of the seismic waves, slope materials and slope sizes).

A dynamic finite element method (FEM) was used to analyze the ground motion process. Even though a three-dimensional (3-D) model was required to fully describe the topographic effects, two-dimensional (2-D) topographic profiles adequately captured the essential geometry of mountain ridges and allowed exploration of key features of the observed landslide patterns [35]. Thus, a 2-D homogeneous double-faced slope with varying slope sizes was established, as illustrated in Figure 1. The height of the slope was assumed to be H, which varied according to the wavelength λ. That is, the height of the slope was considered to be the normalized height in this study [36]. The width of the slope crest, W, corresponded to the slope height. The inclination of slope i evolved from gentle to steep. Moreover, the widths of the ground surfaces were 200 m per side, and the depths of the foundation were 200 m. The seismic waves were input from the left bottom of the foundation with incident angles θ in the range of 0°–30°, with intervals of 5°. To ensure the accurate representation of wave transmissions through a model in the FEM, the element size of the model had to be less than one-twelfth of the wavelength [53], and the time step of the dynamic analysis was determined to be one-tenth of the maximum frequency of the seismic waves [13].

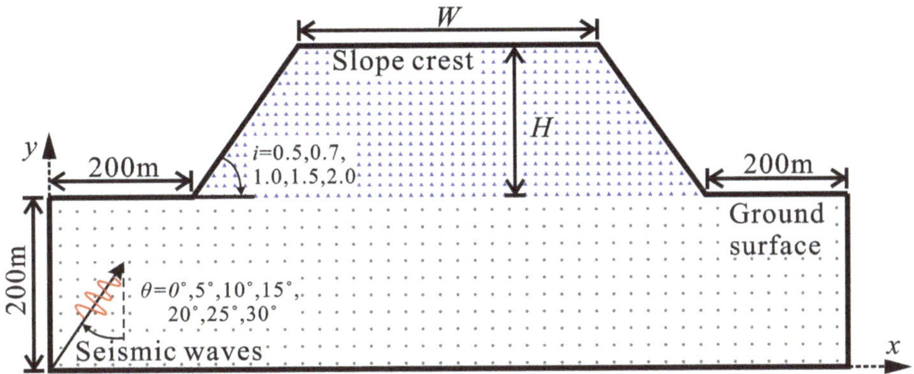

Figure 1. Sketch of the calculated modeling.

The seismic waves were analyzed by two components of the seismic waves (considering only the body waves), that is, P waves and SV waves. The influence of the wave components was analyzed using the same model size. The materials of the slope were assumed to be isotropic, and the medium was linearly elastic. To investigate the influence of materials in the evaluation of the topographic effects, the material of the foundation was fixed, whereas the materials of the slope varied according to the variations in the materials in the foundations. The specific parameters of the materials are presented in Table 1.

Table 1. Parameters of soil medium implemented in FEM.

Soil Types	Mass Density ρ (kg/m³)	Poison's Ratio v	Elastic Modulus E (MPa)	P Wave Velocity c_p (m/s)	S Wave Velocity c_s (m/s)
Foundation	2000	0.2	768	653.1	400
Slope I	2000	0.2	768	653.1	400
Slope II	2000	0.3	425	534.8	285.9
Slope III	2000	0.25	1370	906.4	523.4

3. Input of Seismic Waves in Numerical Simulation

3.1. Establishment of the Artificial Boundary and the Input Method

In a dynamic analysis of the numerical simulation, the study area is usually truncated from an infinite region with an artificial boundary used to absorb the scattered waves. In this study, a viscous-spring artificial boundary was adopted [54,55]. The elastic springs and dampers were established to implement the viscous-spring artificial boundary on the boundary nodes, as shown in Figure 2. The coefficients of the elastic spring (K) and damper (C) are defined as follows.

$$K_N = \frac{1}{1+A} \frac{\lambda + 2G}{R} \cdot A_l, \quad C_N = B\rho c_p \cdot A_l \tag{1}$$

$$K_T = \frac{1}{1+A} \frac{G}{R} \cdot A_l, \quad C_T = B\rho c_s \cdot A_l \tag{2}$$

where A and B are the correction coefficients; the good values of the coefficients are 0.8 and 1.1, respectively [56]; ρ is the mass density; R is the distance from the wave source to the boundary; c_p and c_s are the velocities of the compression wave and shear wave in the medium, respectively; G is the shear modulus; and subscripts T and N indicate the tangential and normal directions, respectively. A_l represents the influence area at each node, for example, at node l, $A_l = (A_1 + A_2)/2$, as depicted in Figure 2.

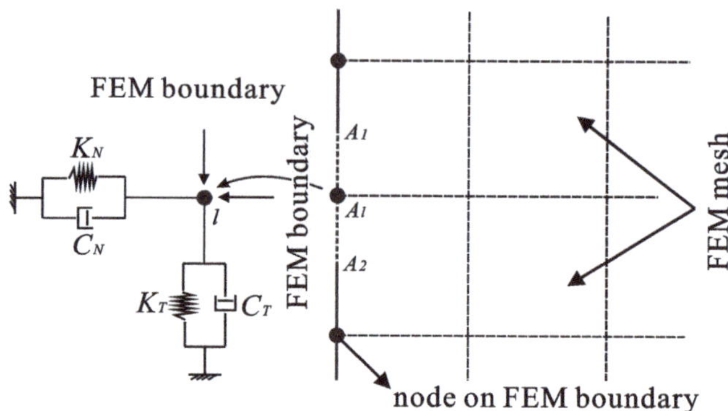

Figure 2. Layout of viscous-spring artificial boundary on FEM model.

The motion equation of the lumped mass in the FEM wavefield at the artificial boundary is expressed as follows:

$$m\ddot{\mathbf{u}} + c\dot{\mathbf{u}} + k\mathbf{u} = A\boldsymbol{\sigma} \tag{3}$$

In case of the distribution of the wavefield on the artificial boundary, the displacement and stress on the artificial boundary can be divided into free-field motions (denoted by superscripts f) and scattered-field motions denoted by superscripts s), which are illustrated as follows:

$$\mathbf{u} = \mathbf{u}^f + \mathbf{u}^s \boldsymbol{\sigma} = \boldsymbol{\sigma}^f + \boldsymbol{\sigma}^s \tag{4}$$

The equation of stress (specifically for the scattered-field motions) on the viscous-spring artificial boundary is given as:

$$\boldsymbol{\sigma}^s = -K\mathbf{u}^s - C\dot{\mathbf{u}}^s \tag{5}$$

Substituting Equations (4) and (5) into Equation (3), the motion equation on the artificial boundary is expressed as follows:

$$m\ddot{\mathbf{u}} + (c + AC)\dot{\mathbf{u}} + (k + AK)\mathbf{u} = A\boldsymbol{\sigma} + K\mathbf{u}^f + C\dot{\mathbf{u}}^f \tag{6}$$

Equation (6) can be considered seismic input and non-radiation, and the right side of the equation represents the equivalent nodal forces. When the two directions of the viscous-spring artificial boundary (normal direction and tangential direction) are considered, the equivalent nodal force can be expressed as follows:

$$\mathbf{f}_{li} = K_{li}\mathbf{u}_{li}^f + C_{li}\dot{\mathbf{u}}_{li}^f + A_l\boldsymbol{\sigma}_{li}^f \tag{7}$$

where subscript i denotes the direction and subscript l denotes the node. K_{li} and C_{li} are the coefficients of the elastic spring and damper, which can be calculated using Equations (1) and (2), respectively; σ_{li}^f, u_{li}^f and $\dot{u}_{li}^f$ represent the tensors of the stress, displacement and velocity, respectively.

3.2. Implementation of Oblique Incident Waves

3.2.1. SV Waves

Based on the artificial boundary, the other key point in the numerical simulation of a dynamic problem is the implementation of seismic waves in finite element models. An efficient and high-precision method is to convert the seismic records, that is, acceleration records, together with the coefficients of the artificial boundary into equivalent nodal forces. Figure 3 presents a plane study area that was subjected to oblique incident SV waves. As shown in Figure 3, each truncated boundary was subjected to three waves that involved one incident wave and two reflected waves. When node A, one of the joints on the left truncated boundary, were considered, the incident seismic waves of node A reached the boundary directly (incident SV waves) or through reflection after first reaching the ground surface (reflected SV waves and reflected P waves). The displacement and velocity of the incident SV waves are denoted as $u_s(t)$ and $\dot{u}_s(t)$, respectively.

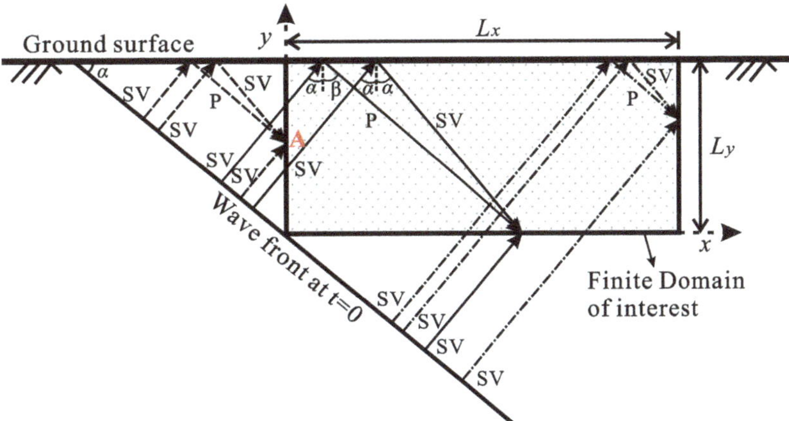

Figure 3. Sketch of the seismic waves on truncated boundary by obliquely incident SV waves.

Consequently, the free-field motion and the associated stress (in terms of x and y components) can be induced at the given point A (x_0, y_0) as follows:

$$
\begin{cases}
u_{lx}(t) = u_s(t - \Delta t_1)\cos(\alpha) - A_1 u_s(t - \Delta t_2)\cos(\alpha) + A_2 u_p(t - \Delta t_3)\sin(\beta) \\
u_{ly}(t) = -u_s(t - \Delta t_1)\sin(\alpha) - A_1 u_s(t - \Delta t_2)\sin(\alpha) + A_2 u_p(t - \Delta t_3)\cos(\beta) \\
\dot{u}_{lx}(t) = \dot{u}_s(t - \Delta t_1)\cos(\alpha) - A_1 \dot{u}_s(t - \Delta t_2)\cos(\alpha) + A_2 \dot{u}_p(t - \Delta t_3)\sin(\beta) \\
\dot{u}_{ly}(t) = -\dot{u}_s(t - \Delta t_1)\sin(\alpha) - A_1 \dot{u}_s(t - \Delta t_2)\sin(\alpha) + A_2 \dot{u}_p(t - \Delta t_3)\cos(\beta)
\end{cases}
\tag{8}
$$

$$
\begin{cases}
\sigma_{lx} = S_1\left(\dot{u}_s(t - \Delta t_1) + S_2 A_1 \dot{u}_s(t - \Delta t_2)\right) + S_3 A_2 \dot{u}_p(t - \Delta t_3) \\
\sigma_{ly} = S_4\left(\dot{u}_s(t - \Delta t_1) + S_5 A_1 \dot{u}_s(t - \Delta t_2)\right) + S_6 A_2 \dot{u}_p(t - \Delta t_3)
\end{cases}
\tag{9}
$$

where A_1 and A_2 are the amplitude ratios (denoted as the reflected SV waves and reflected P waves to the incident SV waves), Δt is the delay time of waves from the wavefront at $t = 0$ to the boundary node (node A), α and β are the angles between the vertical direction and reflected SV waves, and the reflected P waves, respectively. A_1, A_2 and β can be expressed as follows:

$$
\begin{cases}
A_1 = \dfrac{c_s{}^2 \sin 2\alpha \sin 2\beta - c_p{}^2 \cos^2 2\alpha}{c_s{}^2 \sin 2\alpha \sin 2\beta + c_p{}^2 \cos^2 2\alpha} \\[2mm]
A_2 = \dfrac{2 c_p c_s \sin 2\alpha \cos 2\alpha}{c_s{}^2 \sin 2\alpha \sin 2\beta + c_p{}^2 \cos^2 2\alpha} \\[2mm]
\beta = \arcsin\left(\dfrac{c_p \sin \alpha}{c_s}\right)
\end{cases}
\tag{10}
$$

where c_s and c_p are the wave velocities of P waves and SV waves, respectively. The delay times Δt_1, Δt_2, Δt_3, and variables S_1 to S_6 are all boundary-dependent, which can be depicted as follows:

$$\left\{ \begin{array}{l} \Delta t_1 = y \cos \alpha / c_s \\ \Delta t_2 = \left(2L_y - y\right) \cos \alpha / c_s \\ \Delta t_3 = \left(L_y - y\right) / \left(c_p \cos \beta\right) + \left(L_y - \left(L_y - y\right) \tan \alpha \tan \beta\right) \cos \alpha / c_s \end{array} \right.$$
$$S_1 = \frac{G \sin 2\alpha}{c_s}, \ S_2 = -1, \ S_3 = \frac{\lambda + 2G \sin^2 \beta}{c_p},$$
$$S_4 = \frac{G \cos 2\alpha}{c_s}, \ S_5 = 1, \ S_6 = -\frac{G \sin 2\beta}{c_p}$$

$$(11)$$

On the bottom boundary, they became:

$$\left\{ \begin{array}{l} \Delta t_1 = x \sin \alpha / c_s \\ \Delta t_2 = \left(2L_y + x \tan \alpha\right) \cos \alpha / c_s \\ \Delta t_3 = L_y / \left(c_p \cos \beta\right) + \left(L_y \cos \alpha + x \sin \alpha - L_y \tan \beta \sin \alpha\right) / c_s \end{array} \right.$$
$$S_1 = \frac{G \cos 2\alpha}{c_s}, \ S_2 = 1, \ S_3 = -\frac{G \sin 2\beta}{c_p},$$
$$S_4 = \frac{G \sin 2\alpha}{c_s}, \ S_5 = 1, \ S_6 = \frac{\lambda + 2G \cos^2 \beta}{c_p}$$

$$(12)$$

On the right boundary, the stresses were the same but in opposite directions to those on the left boundary, and for the displacement, the delay times were Δt_1, Δt_2, and Δt_3, each of them increased by $L_x \sin \alpha / c_s$ for the additional travel distance.

3.2.2. P Waves

Figure 4 presents a plane study area that was subjected to oblique incident P waves. Similar to the oblique incident SV waves, each truncated boundary was subjected to three waves that involved one incident wave and two reflected waves, as shown in Figure 4. When node B, one of the joints on the left truncated boundary, were considered, the incident seismic waves of node B reached the boundary directly (incident P waves) or through reflection after first reaching the ground surface (reflected P waves and reflected SV waves). The displacement and velocity of the incident P waves are denoted as $u_p(t)$ and $\dot{u}_p(t)$, respectively.

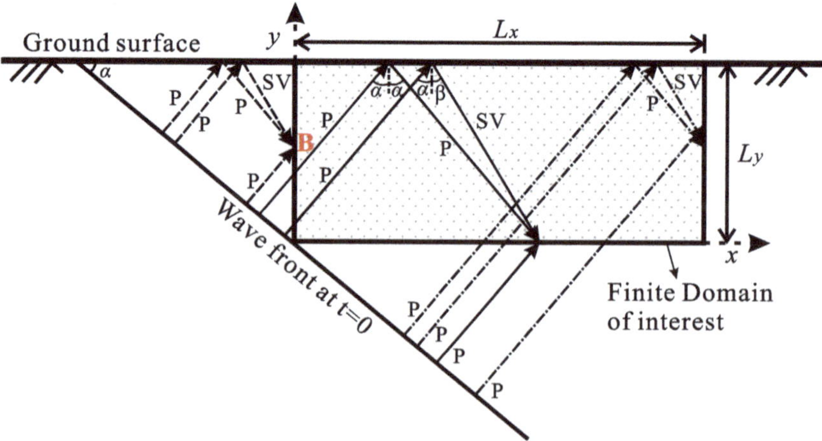

Figure 4. Sketch of the seismic waves on truncated boundary by obliquely incident P waves.

Consequently, the free-field motion and associated stress (in terms of their x and y components) can be induced at the given point B (x_1, y_1) as follows:

$$\left\{ \begin{aligned} u_{lx}(t) &= u_p(t - \Delta t_4)\sin(\alpha) + A_3 u_p(t - \Delta t_5)\sin(\alpha) + A_4 u_s(t - \Delta t_6)\cos(\beta) \\ u_{ly}(t) &= u_p(t - \Delta t_4)\cos(\alpha) - A_3 u_p(t - \Delta t_5)\cos(\alpha) + A_4 u_s(t - \Delta t_6)\sin(\beta) \\ \dot{u}_{lx}(t) &= \dot{u}_p(t - \Delta t_4)\sin(\alpha) + A_3 \dot{u}_p(t - \Delta t_5)sin(\alpha) + A_4 \dot{u}_s(t - \Delta t_6)\cos(\beta) \\ \dot{u}_{ly}(t) &= \dot{u}_p(t - \Delta t_4)\cos(\alpha) - A_3 \dot{u}_p(t - \Delta t_5)cos(\alpha) + A_4 \dot{u}_s(t - \Delta t_6)\sin(\beta) \end{aligned} \right. \tag{13}$$

$$\left\{ \begin{aligned} \sigma_{lx} &= S_7\big(\ddot{u}_p(t - \Delta t_4) + S_8 A_3 \ddot{u}_p(t - \Delta t_5)\big) + S_9 A_4 \ddot{u}_s(t - \Delta t_6) \\ \sigma_{ly} &= S_{10}\big(\ddot{u}_p(t - \Delta t_4) + S_{11} A_3 \ddot{u}_p(t - \Delta t_5)\big) + S_{12} A_4 \ddot{u}_s(t - \Delta t_6) \end{aligned} \right. \tag{14}$$

where A_3 and A_4 are the amplitude ratios (denoted as the reflected P waves and reflected SV waves to the incident P waves). Δt denotes the delay time of waves from the wavefront at $t = 0$ to the boundary node (node B). α and β are the angles between the vertical direction and reflected P waves together with the reflected SV waves, respectively. A_3, A_4 and β can be expressed as follows:

$$\left\{ \begin{aligned} A_3 &= \frac{c_s{}^2 \sin 2\alpha \sin 2\beta - c_p{}^2 \cos^2 2\beta}{c_s{}^2 \sin 2\alpha \sin 2\beta + c_p{}^2 \cos^2 2\beta} \\ A_4 &= \frac{2 c_p c_s \sin 2\alpha \cos 2\beta}{c_s{}^2 \sin 2\alpha \sin 2\beta + c_p{}^2 \cos^2 2\beta} \\ \beta &= \arcsin\left(\frac{c_s \sin \alpha}{c_p}\right) \end{aligned} \right. \tag{15}$$

where c_s and c_p are the wave velocities of P waves and SV waves, respectively. The delay times Δt_4, Δt_5, Δt_6, and the variables S_7 to S_{12} are all boundary-dependent, which can be depicted as follows:

$$\left\{ \begin{aligned} \Delta t_4 &= y \cos \alpha / c_p \\ \Delta t_5 &= \big(2 L_y - y\big)\cos \alpha / c_p \\ \Delta t_6 &= \big(L_y - y\big)/(c_s \cos \beta) + \big(L_y - (L_y - y)\tan \alpha \tan \beta\big)\cos \alpha / c_p \end{aligned} \right. \tag{16}$$
$$S_7 = \frac{\lambda + 2G \sin^2 \alpha}{c_p}, \; S_8 = 1, \; S_9 = \frac{G \sin 2\beta}{c_s},$$
$$S_{10} = \frac{G \sin 2\alpha}{c_p}, \; S_{11} = -1, \; S_{12} = -\frac{G \cos 2\beta}{c_s}$$

On the bottom boundary, they became:

$$\left\{ \begin{aligned} \Delta t_4 &= x \sin \alpha / c_p \\ \Delta t_5 &= \big(2 L_y + x \tan \alpha\big)\cos \alpha / c_p \\ \Delta t_6 &= L_y/(c_s \cos \beta) + \big(L_y \cos \alpha + x \sin \alpha - L_y \tan \beta \sin \alpha\big)/c_p \end{aligned} \right. \tag{17}$$
$$S_7 = \frac{G \sin 2\alpha}{c_p}, \; S_8 = -1, \; S_9 = -\frac{G \cos 2\beta}{c_s},$$
$$S_{10} = \frac{\lambda + 2G \cos^2 \alpha}{c_p}, \; S_{11} = 1, \; S_{12} = -\frac{G \sin 2\beta}{c_s}$$

On the right boundary, the stresses were the same but in directions opposite to those on the left boundary, and for the displacement, the delay times were Δt_4, Δt_5, and Δt_6, each of which increased by $L_x \sin \alpha / c_p$ for the additional travel distance.

3.3. Verification

The incident seismic waves were implemented using the ABAQUS software, together with a self-developed MATLAB program. The specific implementation is described in the flowchart illustrated in Figure 5. Two numerical test examples were used to verify the validity of the oblique incident seismic waves. One was the propagating process of oblique incident waves in a truncated region, which was considered to explore the accuracy of the input method in a semi-infinite field. The other was the response of a semicircular canyon input by oblique incident waves, which was aimed at determining the accuracy of the simulation in the topography.

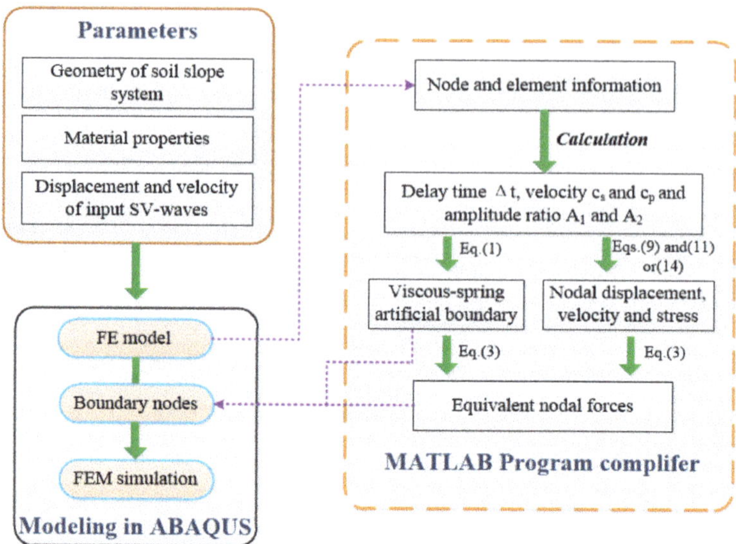

Figure 5. Flow chart for input seismic waves implemented in ABAQUS.

3.3.1. Test Example 1

A truncated rectangular domain was adopted to simulate the propagation process of oblique incident waves in a semi-infinite field, as depicted in Figure 6a. The sizes of the truncated regions were as follows: $L_x = 800$ m and $L_y = 400$ m. The region was sufficiently large to detect the propagation of all the incident and reflected waves. The incident angles of the P waves and SV waves were assumed to be 30° and 18°, respectively.

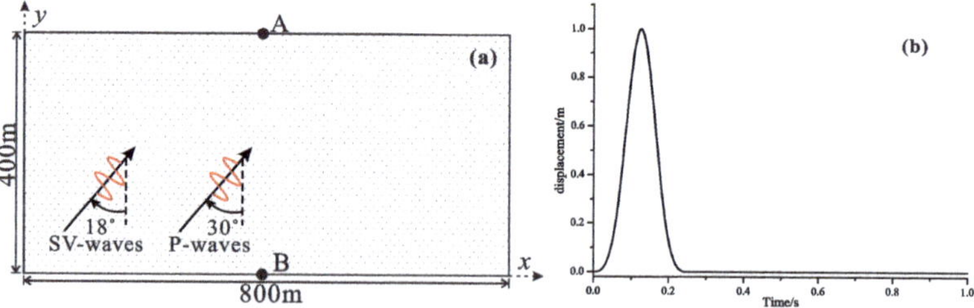

Figure 6. Numerical model and incident waves of test example 1: (**a**) finite element model and (**b**) displacement time history of incident waves.

The material parameters of the domain were as follows: Poisson's ratio = 0.25, elastic modulus = 1.25 Gpa and mass density, $\rho = 2000$ kg/m^3. Points A (400, 400) and B (400, 0) are the observation points. An impulse was adopted as an incident seismic wave, as shown in Figure 6b. The impulse equation is defined as follows [57]:

$$P(t) = 16P_0\left[G(t) - 4G\left(t - \frac{1}{4}\right) + 6G\left(t - \frac{1}{2}\right) - 4G\left(t - \frac{3}{4}\right) + G(t - 1)\right] \qquad (18)$$

where $G = (t/T_0)^3 H(t/T_0)$ $H(t)$ is the Heaviside function, P_0 is the peak value of an impulse, P_0 is set to 1.0 m; and T_0 is the acting time of the impulse, where $T_0 = 0.25$ s.

Figure 7 shows the displacement contours with the time of incident SV waves. As illustrated in the figures, the propagation processes of the incident waves and reflected waves at the ground surface were simulated effectively in a semi-infinite field. In addition, the displacement time histories of the observation points (points A and B) are shown in Figure 8. As shown in the figures, the numerical results proposed by the input method were in good agreement with the analytical solutions, where the analytical solutions were calculated using the elastic wave propagation theory [58,59].

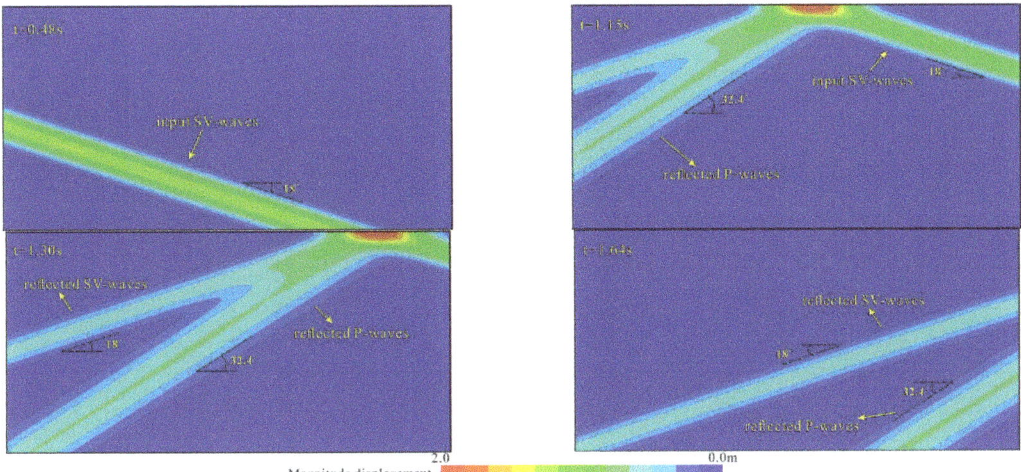

Figure 7. Displacement contours of SV waves propagated in semi-infinite space.

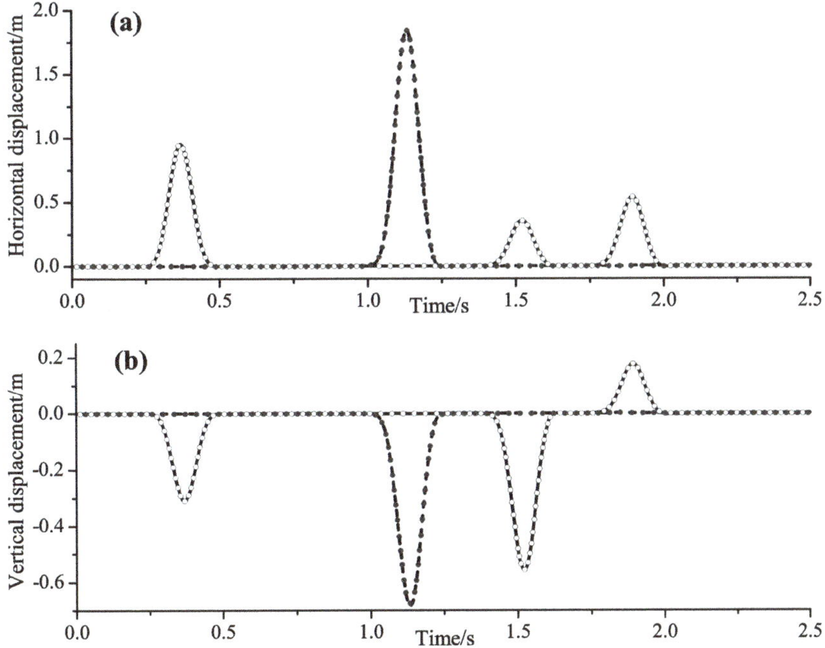

Figure 8. Displacement time history at observation points of incident SV waves: (**a**) horizontal displacement, and (**b**) vertical displacement (solid lines and dashed lines are numerical results of points A and B, respectively, with circles showing analytical solutions).

The displacement contours with the time of P waves in Figure 9 show the propagation processes of the incident waves and reflected waves at the ground surface. Meanwhile, the displacement contours and the displacement time histories of the observation points (points A and B) in Figure 10 both confirm that the input method was effective and precise in the semi-infinite field. Therefore, the oblique incident SV waves and P waves in a semi-infinite field could be simulated using the proposed input method, and the ability of the viscous-spring artificial boundary to absorb the scattered waves was also verified.

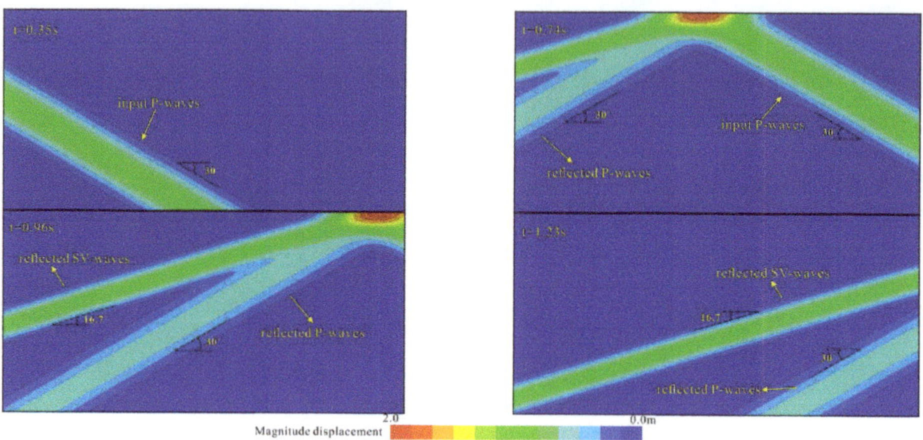

Figure 9. Displacement contours of P waves propagated in semi-infinite space.

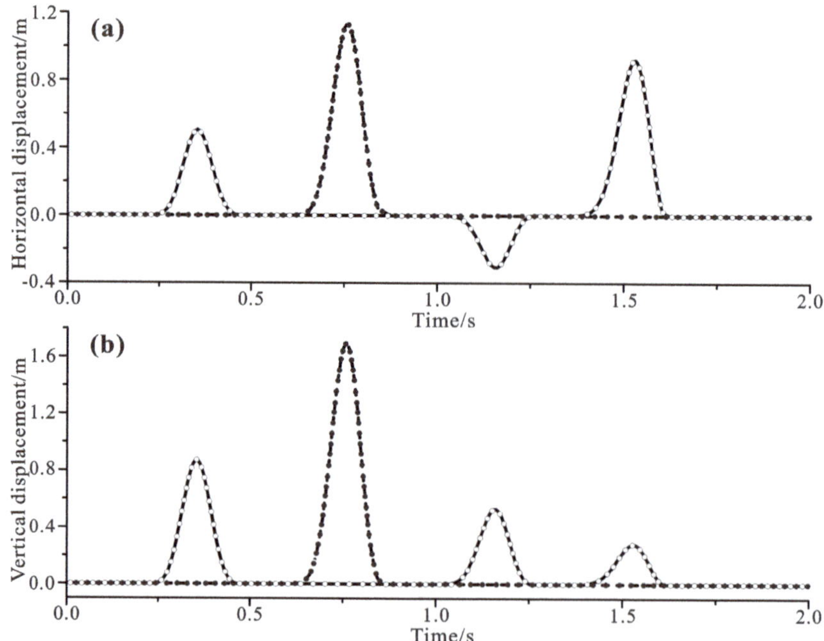

Figure 10. Displacement time history at observation points of incident P waves: (**a**) horizontal displacement, and (**b**) vertical displacement (solid lines and dashed lines are numerical results of points A and B, respectively, with circles showing analytical solutions).

3.3.2. Test Example 2

A truncated rectangular domain with a semicircular canyon is shown in Figure 11a. The truncated region was $0 \leq x \leq 200$ m, $0 \leq y \leq 200$ m and the radius of the canyon was 25 m. The material parameters of the ground were as follows: Poisson's ratio = 0.3, elastic modulus = 208 MPa and mass density, ρ = 2000 kg/m^3. A Ricker wavelet was adopted as the incident wave. The acceleration-time history of the wave is illustrated in Figure 11b. The time history of the Ricker wavelet was defined as follows:

$$R(t) = \left(1 - 2\pi^2 f_0^2 t^2\right) \exp\left(-\pi^2 f_0^2 t^2\right) \tag{19}$$

with an excitation frequency f_0 of 4.0 Hz.

The incident angles had values of 25° under SV waves and 30° under P waves. Figures 12 and 13 show the horizontal displacement amplification ratios U_h and vertical displacement amplification ratios U_v of the ground surface on the canyon with incident SV waves and incident P waves, respectively. U_h is defined as the ratio of the horizontal displacement on the ground surface to the input horizontal displacement, and U_v is the ratio of the vertical displacement. The numerical results presented in Figures 12 and 13 were in good agreement with the analytical solutions proposed by Wong [60]. Therefore, in the topographic analysis, the proposed input method could effectively simulate the propagation process of oblique incident seismic waves, and the viscous-spring artificial boundary could absorb the scattered waves perfectly because of the topography.

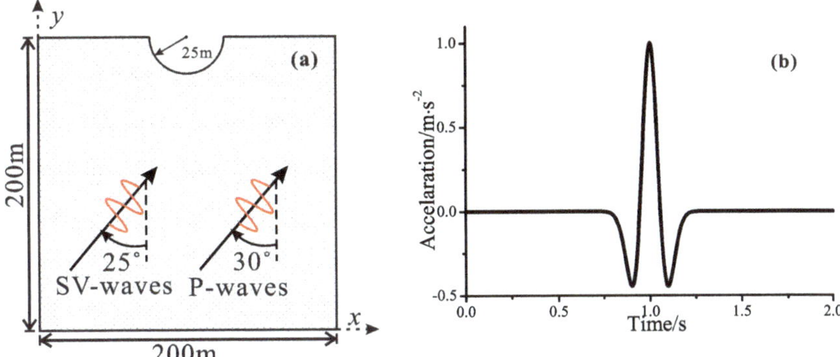

Figure 11. Numerical model and input waves of test example 2: (**a**) finite element model and (**b**) acceleration time-history of incident waves.

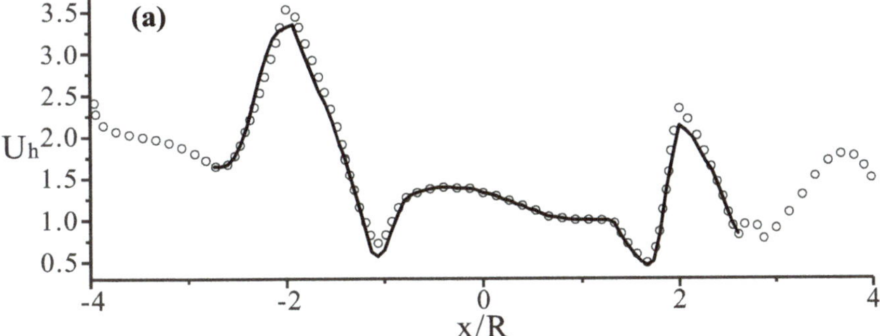

Figure 12. *Cont.*

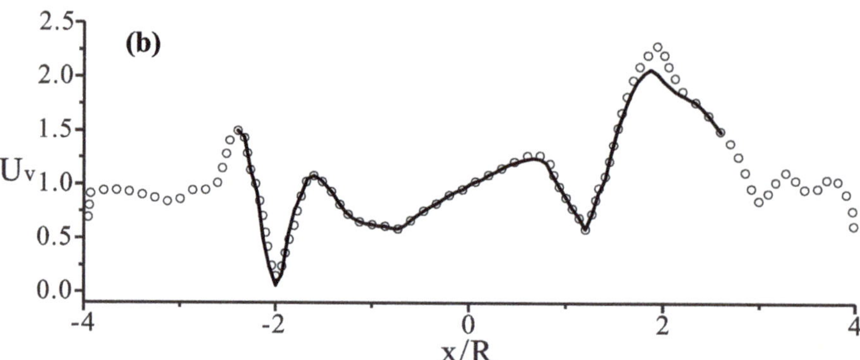

Figure 12. Displacement amplification ratios of ground under incident SV-waves with angle of 25°: (**a**) horizontal ratios and (**b**) vertical ratios (solid lines are analytical solutions by Wong [60] and circles are numerical results of finite element method based on ABAQUS).

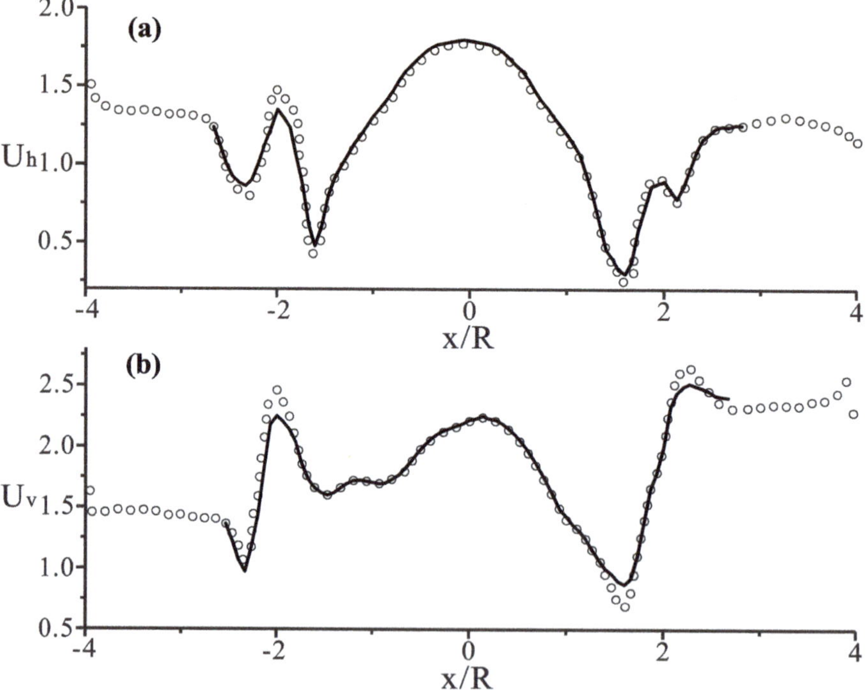

Figure 13. Displacement amplification ratios of ground under incident P-waves with angle of 30°: (**a**) horizontal ratios and (**b**) vertical ratios (solid lines are analytical solutions by Wong [60] and circles are numerical results of finite element method based on ABAQUS).

4. Discussion and Results

A schematic of the numerical model is illustrated in Figure 1. The mesh type of this numerical model is quadrilateral element. A Ricker wavelet with a predominant frequency of 4.0 Hz was employed as the incident wave to investigate the wave patterns, materials and slope geometries based on the seismic responses. The acceleration-time history of the incident wave, which was input from the left bottom of the model, is illustrated in Figure 10b. To analyze the topographic effects, the acceleration amplification ratio was applied, which is the ratio of the peak acceleration of the reflected waves at each point on the surface to that of the incident seismic waves [36,45]. The horizontal and vertical acceleration amplification ratios r_h and r_v are defined in Equations (20) and (21), respectively. For a given wave, r_h and r_v are the values of the acceleration amplification ratio r for the horizontal and vertical components, respectively. $r = \sqrt{r_h^2 + r_v^2}$ defines zones of the net amplification of ground acceleration with respect to the energy carried by the incoming waves.

$$r_h = \frac{\max(|a_h|)}{|a_{h,input}|} \tag{20}$$

$$r_v = \frac{\max(|a_v|)}{|a_{v,input}|} \tag{21}$$

where $\max(|a_h|)$ and $\max(|a_v|)$ are the peak horizontal (h) and peak vertical (v) acceleration at the observation points along the slope ridges and slope crests, respectively; and $|a_{h,input}|$ and $|a_{v,input}|$ are the horizontal (h) and vertical (v) acceleration, respectively.

In this study, the wave patterns were discussed based on the P waves and the SV waves owing to the specific focus on the effects of body waves. In addition, the impact of the materials was investigated in terms of the relative hardness and softness between the slopes and foundations. The shear waves of different materials on the slope topographies are presented in Table 1. Furthermore, the geometries of slopes with varying slope heights, widths and inclinations are discussed. The slope height H is a normalized height that is related to the wavelength λ [29,36], and it varies among 0.2λ, 0.5λ, 1.0λ and 2.0λ. The width of the slope crest W varied from 50 to 400 m, and the intermediate values were 100 m and 200 m. The slope inclination i varied between $26.6°$, $33.7°$, $45°$, $55°$ and $63.4°$, corresponding to width-depth ratios (defined as $\cot i$) of 2.0, 1.5, 1.0, 0.7 and 0.5. All the influencing factors (wave patterns, materials and geometries) were based on the varied directions of incident waves, which were input from the left bottom of the topography with incident angles θ in the range of $0°$ to $30°$, sampled at $5°$ intervals.

4.1. Effects of Wave Patterns

Wave patterns, that is, P waves and SV waves, were discussed based on three different slope topographies with varied angles of incident waves. Three models with different slope widths, $W = 0$ m, 100 m and 400 m, were built to evaluate the effect of wave patterns on ground motion amplification. The original model had a height of 100 m and an inclination of 1.0 ($45°$); meanwhile, the materials of the slope were regarded as homogeneous. The acceleration amplification ratios of the P waves and SV waves at different seismic wave inclinations are summarized in Figure 14.

The acceleration amplification ratios of SV waves were larger than those of P waves at a specific slope topography. The slope width became wider and the amplification ratios gradually decreased for each slope pattern. With an increase in the angle of incidence, the acceleration amplification effects moved away from the epicenter and the maximum acceleration amplification ratios were obtained when the incident angles fluctuated between $20°$ and $30°$, which were especially clear under SV waves. Thus, the pattern of amplification of the SV waves was verified to be stronger in the slope topography, and the evolution was the same as that of the findings of Meunier [35] (depicted as the slope configurations of Figure 14a).

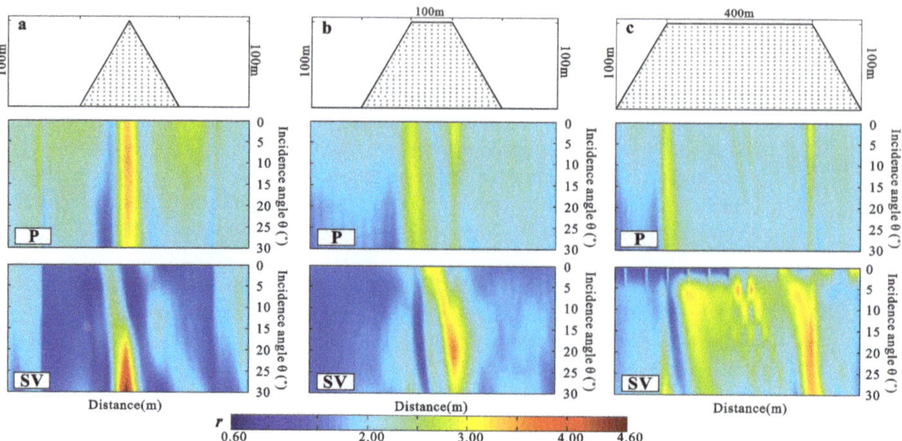

Figure 14. Variations of the seismic responses along the slope ridges of the real amplification ratio (*r*) of PGA with varied angles of wave incidence at P waves and SV waves. (**a**) slope width of 0 m, (**b**) slope width of 100 m, (**c**) slope width of 400 m.

4.2. Effects of Materials

In this section, materials are discussed based on the specific slope topography with varying directions of incident waves. Three models, wherein the slope is soft whereas the foundation is hard, the slope and foundation are both the same, and the slope is hard whereas the foundation is soft, were built to investigate the effects of the materials on the ground motion amplification of slope topography. The investigated model was 100 m high, 100 m wide and 1.0 (45°) in inclination. The so-called softness and hardness of the materials in the slopes or foundations in Figure 15 is in contrast to that of the materials in the foundation (the foundation materials are fixed); that is, the materials of the slope changed from soft to hard (from left to right in Figure 15). The acceleration amplification ratios of the different materials at different incident angles are summarized in Figure 15.

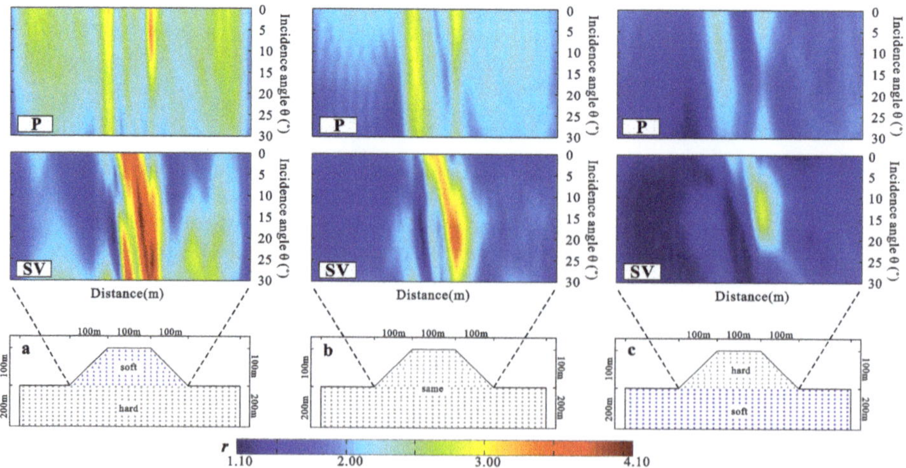

Figure 15. Variations of the seismic responses along the slope ridges and the slope crests of the real amplification ratio (*r*) of PGA with varied angles of wave incidence under different materials. (**a**) slope soft while foundation hard, (**b**) slope and foundation are same, (**c**) slope hard while foundation soft.

In Figure 15, the acceleration amplification ratios of the SV waves were larger than those of the P waves for each material. The ratios gradually decreased as the materials in the slope varied from soft to hard; that is, the soft materials intensified the amplification whereas the hard materials alleviated the amplification in the slope topography. In addition, the maximum acceleration amplification ratios were acquired when the incident angles fluctuated between 10° and 25° under SV waves incidence and fell between 0° and 10° under P waves incidence. Consequently, the slope topographies that covered soft materials were more harmful during earthquake events, and the maximum ground motions of slope topographies were usually in a certain direction of oblique incident seismic waves.

The synthetic accelerograms of the horizontal and vertical components of SV waves and P waves with incident angles of 0° and 30° are presented in Figures 16 and 17, respectively. In the figures, "soft-hard" was the abbreviation of "the slope is soft and the foundation is hard," which corresponded to the situation of "a" in Figure 15; "same" referred to the situation of "b" in Figure 15, and "hard-soft" was marked as the situation of "c" in Figure 15.

As illustrated in the figures, the soft materials led to intense amplification of the ground motions, whereas the hard materials caused de-amplification. The scattered waves under the soft materials were abundant, whereas they were weakened in the hard materials. These evolutions were observed in the horizontal components of the SV waves and the vertical components of the P waves. In addition, compared with the scattered waves at 0° incidence and 30° incidence in Figures 16 and 17, the scattered waves were much more abundant at the oblique incident waves owing to the added refraction effects by the oblique directions of incident waves. In summary, the intense ground motions with soft materials in slope topography could be explained by the abundantly scattered waves, and this phenomenon would aggravate at oblique incident waves.

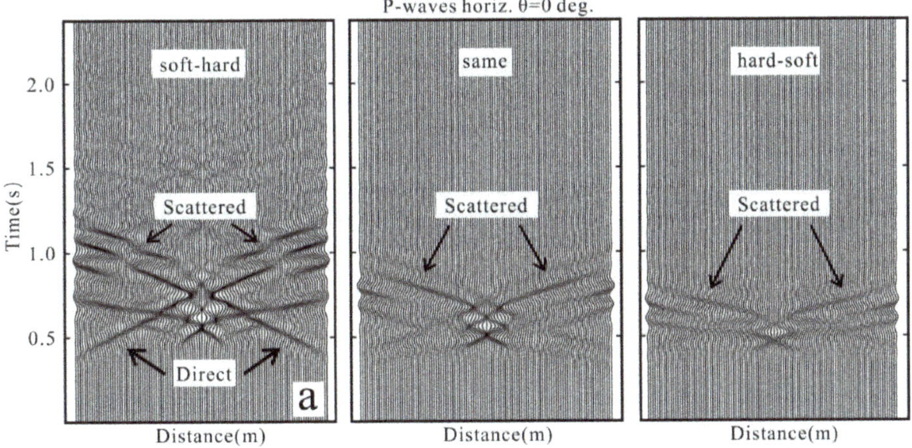

Figure 16. *Cont.*

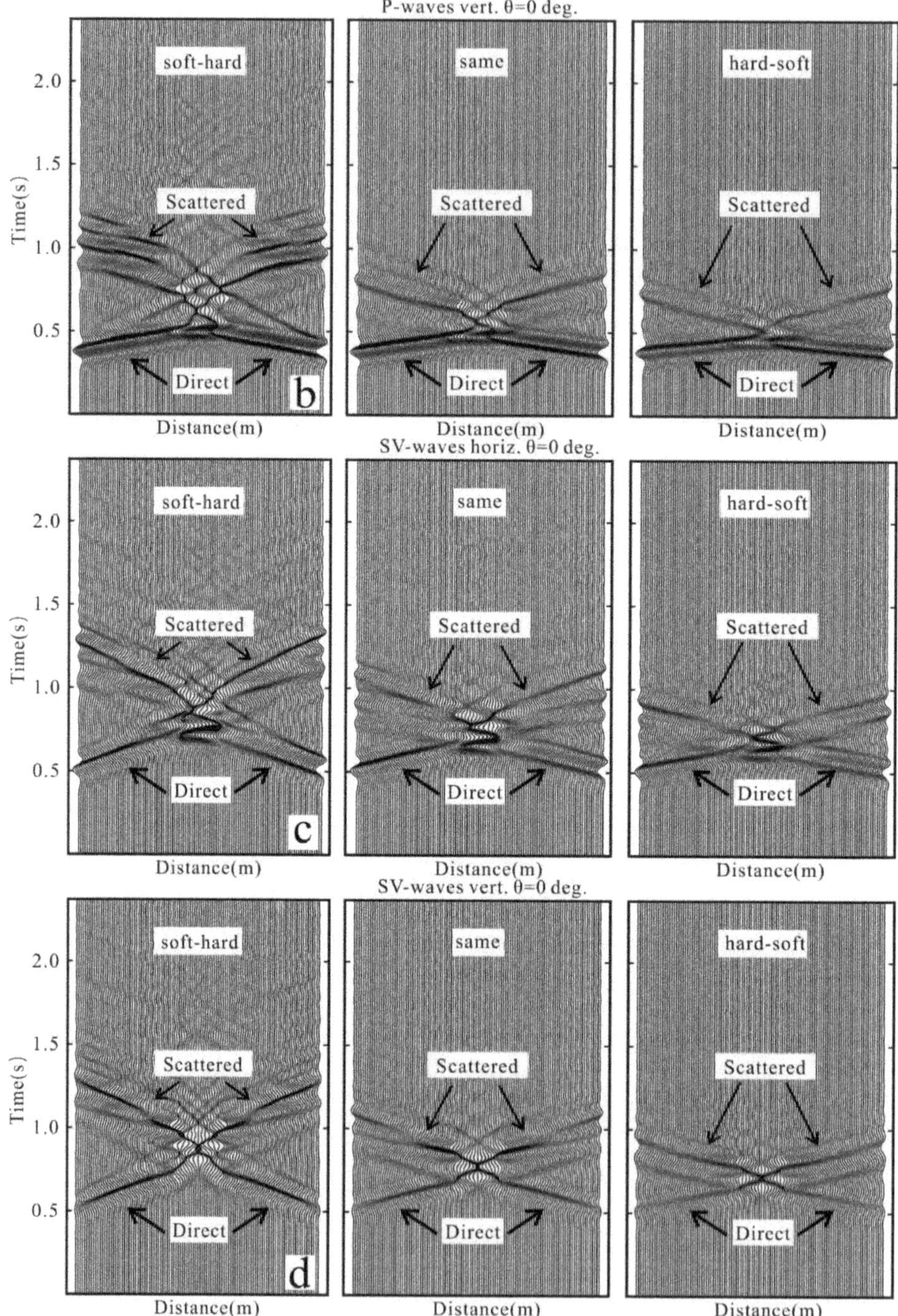

Figure 16. Synthetic accelerogram (accelerations recorded along the ground with time) generated along the topographic profile, plotted from the horizontal or vertical components of P waves and SV waves arriving from the left of the model at an incident angle of 0, (**a**) horizontal P waves with vertical direction, (**b**) vertical P waves with vertical direction, (**c**) horizontal SV waves with vertical direction (**d**) vertical SV waves with vertical direction.

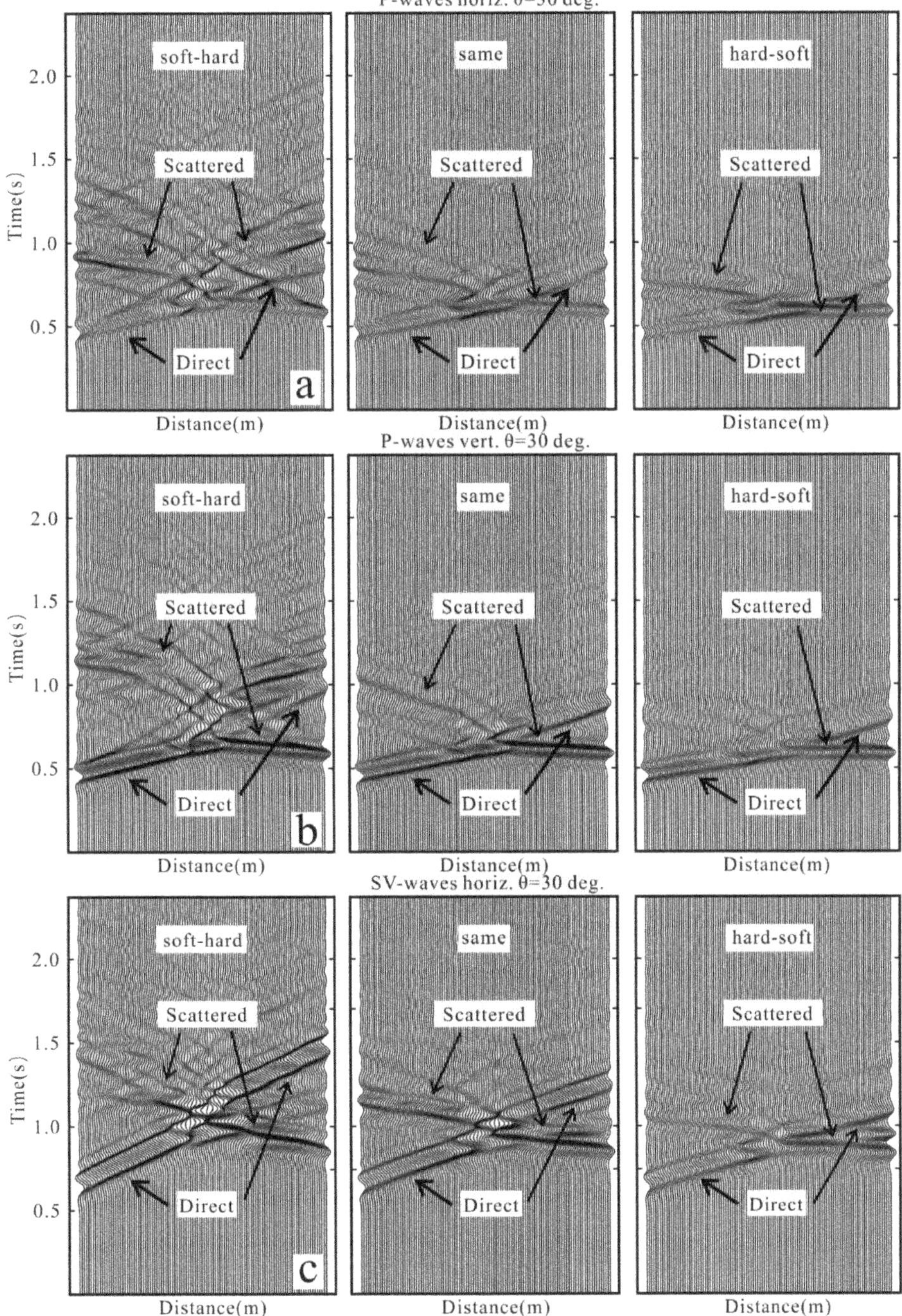

Figure 17. *Cont.*

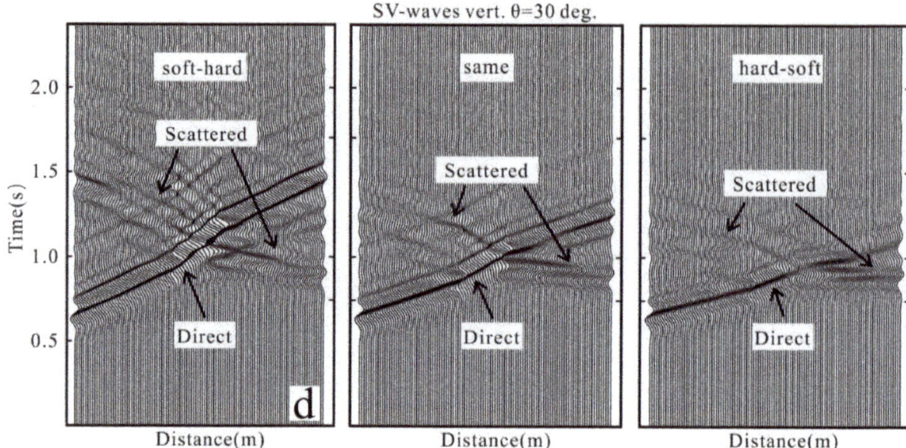

Figure 17. Synthetic accelerogram (accelerations recorded along the ground with time) generated along the topographic profile, plotted from the horizontal or vertical components of P waves and SV waves arriving from the left of the model at incident angle of 30°. (**a**) horizontal P waves with 30° directions, (**b**) vertical P waves with 30° directions, (**c**) horizontal SV waves with 30° directions (**d**) vertical SV waves with 30° directions.

4.3. Effects of Slope Geometries

According to the analysis of the impacts of wave patterns, the pattern of P waves was weak because of the amplification ratios of the ground motions, whereas that of SV waves was strong. Thus, the effects of the slope geometries were discussed only based on the SV waves rather than the P waves. In addition, the slope geometry changed the wave paths at various directions of wave incidence; thus, the materials of the slope topographies were assumed to be homogenous to eliminate the impacts of non-geometric factors. Consequently, this section, discusses the investigation of the effects of the slope geometries using homogeneous materials under SV incident waves.

4.3.1. Influence of Slope Height H

In this subsection, the slope height varied from 20 m, 50 m, 100 m, to 200 m, whereas the other geometrical parameters (slope widths and slope inclinations) remained the same. The acceleration amplification ratios along the slope surface are illustrated in Figure 18 with varying angles of incident waves.

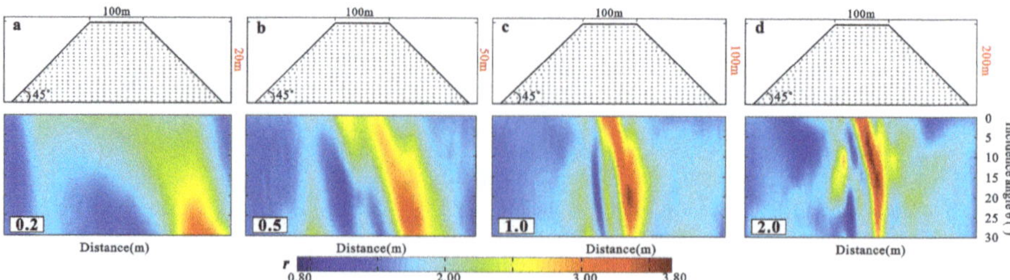

Figure 18. Variations of the seismic responses along the slope ridges and the slope crests of the real amplification ratio (*r*) of PGA with varied angles of wave incidence at different slope heights. (**a**) slope height of 20 m, (**b**) slope height of 50 m, (**c**) slope height of 100 m, (**d**) slope height of 200 m.

In Figure 18, different distributions of ground motions are presented with varying slope heights under SV incident waves. The acceleration amplification ratios rose significantly with an increase in the slope height from 0.2λ to 2.0λ. The area of the maximum acceleration amplification ratios at each height moved to the center of the slope topography with an increase in slope height. In addition, the maximum acceleration amplification ratios could be obtained when the incident angles fell between $25°$ and $30°$ under a slope height of 0.2λ, whereas the maximum acceleration amplification ratios could be obtained when the incident angles fluctuated between $5°$ and $10°$ under a slope height of 2.0λ. Thus, the higher the slope heights, the closer the maximum acceleration amplification ratios were to the smaller inclinations of wave incidence. In summary, different slope heights influenced the wave aggregation at the crest; the higher the model, the larger the acceleration amplification ratios. The maximum ground motions were obtained in the oblique direction of the incident waves.

Moreover, the observation points (slope toe and slope crest) are presented with horizontal and vertical components in Figures 19 and 20, respectively, to analyze the comprehensive relations between the incident directions and slope heights.

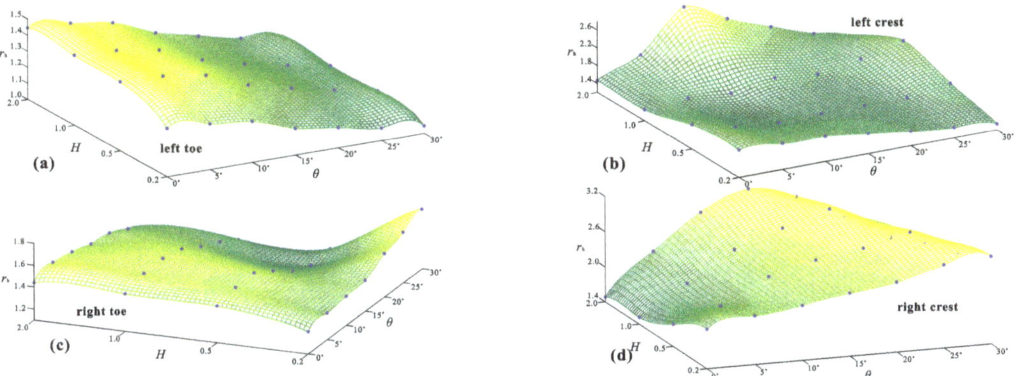

Figure 19. Variations of the horizontal amplification ratios (r_h) of PGA at the observation points ((**a**) left toe, (**b**) left crest, (**c**) right toe and (**d**) right crest) with varied angles of wave incidence at different slope heights.

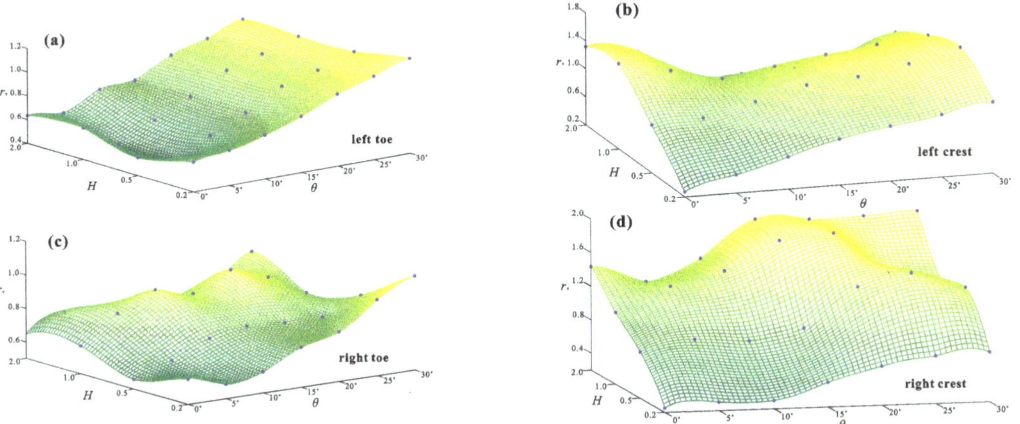

Figure 20. Variations of the vertical amplification ratios (r_v) of PGA at the observation points ((**a**) left toe, (**b**) left crest, (**c**) right toe and (**d**) right crest) with varied angles of wave incidence at different slope heights.

As shown in Figures 19 and 20, the variations in the amplification ratios at the slope toe were gentle. The horizontal amplification ratios at the left toe decreased with an increase in the angle of incidence and increased with an increase in slope height. However, the vertical amplification ratios at the left toe increased with an increase in the angle of incidence, whereas their variations remained almost unchanged with varying slope heights. The horizontal amplification ratios at the right toe increased with an increase in the angle of wave incidence (at a slope height of 0.2), whereas they first increased and then decreased as the incident angle increased (at slope heights of 0.5, 1.0 and 2.0). The vertical amplification ratios of the right toe grew with an increase in the angle of incidence. Nevertheless, these ratios increased significantly at angles of 10° and 20° (at a slope height of 1.0). The variations above were in good agreement with the seismic responses of the ground motions in Figure 18 (at a slope height of 1.0), which indicated that the variations in the horizontal and vertical amplification ratios were non-monotonous at the slope toe.

The variations in the horizontal and vertical amplification ratios at the slope crest are complicated. The horizontal amplification ratios at the left crest monotonously rose or decreased with an increase in the angle of incidence (at slope heights of 0.2, 0.5 and 1.0). However, the horizontal ratios first increased and then decreased with an increase in the angles of incidence (at a slope height of 2.0). The maximum horizontal amplification ratios fell between 5° and 15° (at a slope height of 2.0). The horizontal amplification ratios at the right crest first increased and then decreased as the incident angle increased, and the maximum horizontal amplification ratios moved from the an incident angle of 30° to those at 15° as the slope height increased. The vertical amplification ratios of the left crest increased with an increase in the angle of incidence, and the maximum vertical amplification ratios of the left crest were obtained at a slope height of 1.0. The vertical amplification ratios of the right crest increased with an increase in the angle of incidence (at slope heights of 0.2 and 0.5). The variations in the vertical amplification ratios of the right crest were intensely complicated (at slope heights of 1.0 and 2.0), and the maximum ratios were obtained when the incident angles fell between 10° and 20°.

In summary, the variations in the amplification ratios at the slope toe were monotonous, that is, the ground motions changed regularly with varying incident angles and heights. Nonetheless, the variations in ground motions at the slope crest were complicated owing to the interactions between the input waves and the reflected waves in the topography.

4.3.2. Influence of the Width of Slope Crest W

The width of the slope crest varied from 50 m, 100 m and 200 m to 400 m, whereas the other geometrical parameters (slope heights and slope inclinations) remained the same. The acceleration amplification ratios along the slope surface are illustrated in Figure 21 with varying incident angles.

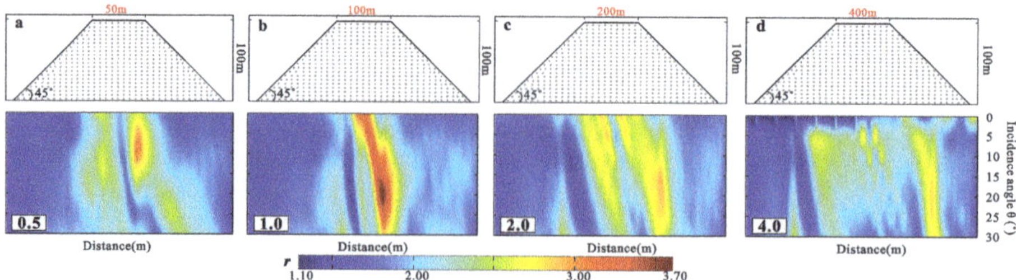

Figure 21. Variations of the seismic responses along the slope ridges and slope crests of the real amplification ratios (*r*) of PGA with varied angles of wave incidence at different slope widths, (**a**) slope width of 50 m, (**b**) slope width of 100 m, (**c**) slope width of 200 m (**d**) slope width of 400 m.

As illustrated in Figure 21, the acceleration amplification ratios at a slope width of 1.0 H were the greatest. This phenomenon can be indicated by the resonance of the model sizes and wavelengths. The ground motions were concentrated on the slope crest, especially when the slope width was 4.0 H. In addition, the maximum acceleration amplification ratios were obtained when the incident angles were between 5° and 10° at a width of 0.5 H, whereas the ratios reached a maximum when the incident angles fluctuated between 25° and 30° at a slope width of 4.0 H. Thus, the wider the slope crests, the closer the maximum acceleration amplification ratios were to the greater inclinations of incident waves. In summary, the energies of earthquake waves were mainly concentrated at the crest of the slope topography. The maximum ground motions occurred as the width of the slope model was nearly one wavelength, especially in certain directions of seismic incident waves.

Moreover, the observation points (slope toe and slope crest) are analyzed in Figures 22 and 23, respectively. All the observation points are illustrated based on the horizontal and vertical amplification ratios, to analyze the comprehensive relations between the incident angles and slope widths.

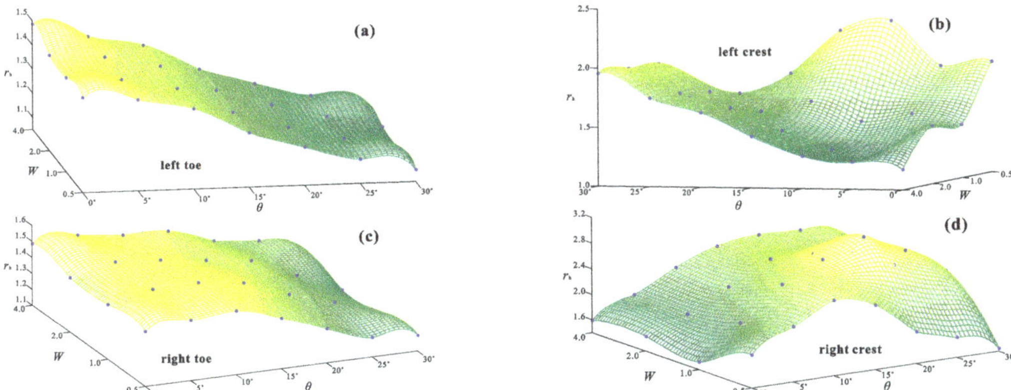

Figure 22. Variations of the horizontal amplification ratios (r_h) of PGA at the observation points ((**a**) left toe, (**b**) left crest, (**c**) right toe and (**d**) right crest) with varied angles of wave incidence at different slope widths.

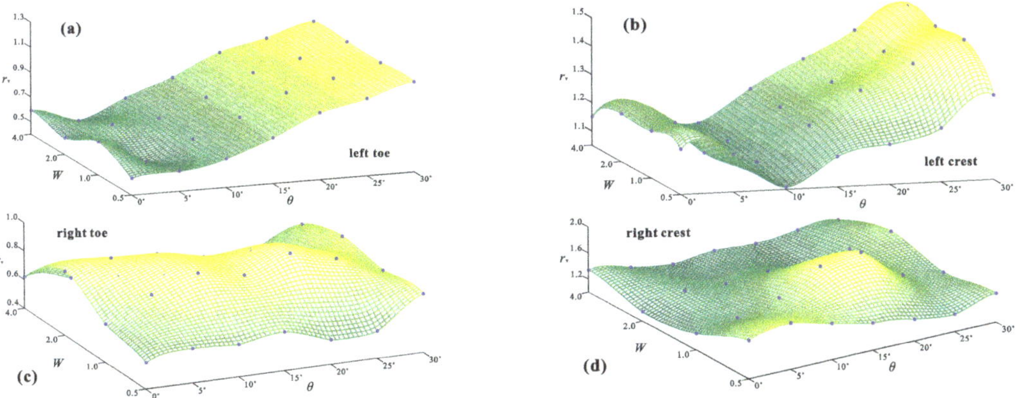

Figure 23. Variations of the vertical amplification ratios (r_v) of PGA at the observation points ((**a**) left toe, (**b**) left crest, (**c**) right toe and (**d**) right crest) with varied angles of wave incidence at different slope widths.

The horizontal amplification ratios at the slope toe decreased monotonously with an increase in the incident angle, whereas they increased as the slope width increased. The vertical amplification ratios at the left toe increased with an increase in the incident angle.

The vertical amplification ratios at the right toe reached a maximum when the slope widths were between 1.0 and 2.0. With an increase in the incident angle, the vertical amplification ratios were almost the same for each slope width.

Nonetheless, variations in the acceleration amplification ratios at the slope crest were complicated. The horizontal amplification ratios at the left crest first decreased and then increased with an increase in the incident angle (at slope widths of 2.0 and 4.0). The horizontal ratios first increased and then decreased as the incident angle increased (at slope widths of 0.5 and 1.0). In addition, the ratios increased from an incident angle of 5° to an incident angle of 20° at a slope width of 0.5, which could be interpreted by the superposition of the scattered waves in the narrow models. The vertical amplification ratios at the left crest first decreased and then increased with an increase in the angle of incidence. Furthermore, the horizontal amplification ratios at the right crest first increased and then decreased with an increase in the angle of incidence. The maximum horizontal ratios were obtained between 15° and 25° of the incident angle (at a slope width of 1.0). The vertical amplification ratios at the right crest were almost the same for various incident angles. The vertical ratios reached a maximum between 15° and 25° of the incident angles (at a slope width of 1.0). The above variations were in good agreement with the seismic responses of ground motions of PGA in Figure 21b, and these evolutions could be interpreted by the correlations between the wavelength and the slope widths.

In summary, the variations in the amplification ratios were mainly concentrated on the slope crest with varying angles of incidence, and these complexities were aggravated when the width was almost one wavelength.

4.3.3. Influence of Slope Inclination I

The inclination of the slope, which is defined as the width-depth ratio, varied from 0.5, 0.7, 1.0 and 1.5 to 2.0, whereas the other geometrical parameters (slope heights and slope widths) remained the same. The acceleration amplification ratios along the slope surface are illustrated in Figure 24 with varying angles of incidence.

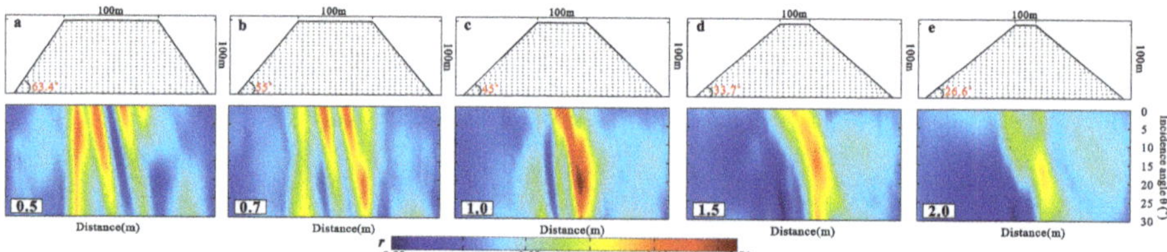

Figure 24. Variations of the seismic responses along the slope ridges and slope crests of the real amplification ratios (r) of PGA with varied angles of wave incidence at different slope inclinations. (**a**) slope inclination of 0.5, (**b**) slope inclination of 0.7, (**c**) slope inclination of 1.0 (**d**) slope inclination of 1.5 (**e**) slope inclination of 2.0.

In Figure 24, the acceleration amplification ratios moved from the left slope crest to the right slope crest when the slope model became gentle. The amplification ratios moved away from the wave source with an increase in the angles of incidence, especially in the gentle slope model. These evolutions are indicated by the scattered waves induced by slope inclinations. The ground motions reached a maximum when the incident angles fluctuated between 5° and 10° in the steep slope model ($i < 1.0$). However, the ground motions reached a maximum when the incident angles were between 15° and 20° in the gentle slope model ($i > 1.0$). Thus, the greater the slope inclinations, the closer the maximum acceleration amplification ratios were to the greater inclinations of the incident waves. In summary, the ground motions were mainly concentrated on the slope ridges. The oblique incident waves and slope inclinations changed the propagation of the seismic wave paths, and the scattered waves easily concentrated as the slope became gentle. The maximum acceleration amplification ratios were obtained in the oblique direction of the incident waves.

Similar to the analyses of the slope height and slope widths, the slope toe and slope crest are also presented with horizontal amplification ratios and vertical amplification ratios in Figures 25 and 26, respectively. The comprehensive relations between the incident angles and slope inclinations are discussed.

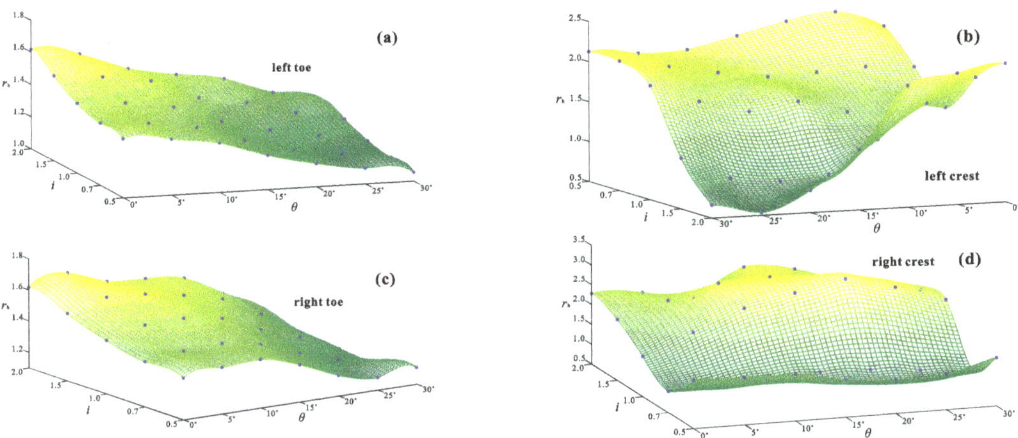

Figure 25. Variations of the horizontal amplification ratios (r_h) of PGA at the observation points ((**a**) left toe, (**b**) left crest, (**c**) right toe and (**d**) right crest) with varied angles of wave incidence at different slope inclinations.

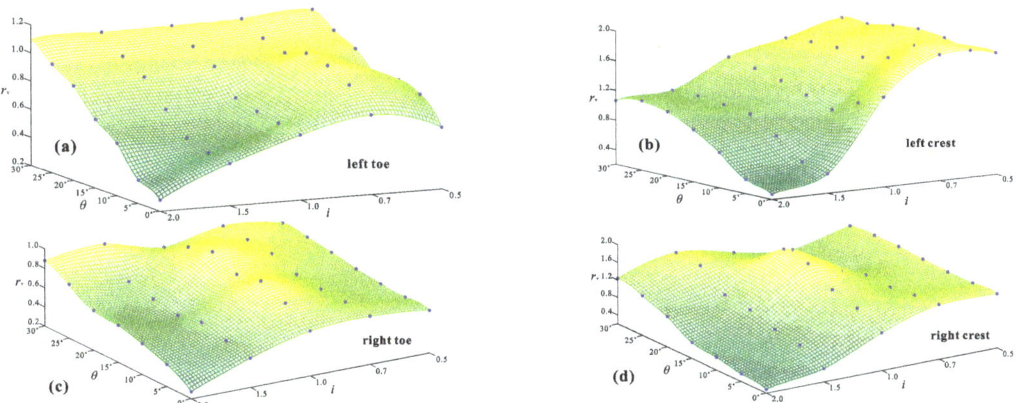

Figure 26. Variations of the vertical amplification ratios (r_v) of PGA at the observation points ((**a**) left toe, (**b**) left crest, (**c**) right toe and (**d**) right crest) with varied angles of wave incidence at different slope inclinations.

The horizontal amplification ratios of the slope toe monotonously decreased with an increase in the angle of incidence, whereas the horizontal ratios increased as the slope inclination increased. That is, the ground motions at the slope toe increased as the slope became gentle. However, the vertical amplification ratios of the slope toe increased monotonously with an increase in the angle of incidence, and the ratios gradually decreased when the slope became gentle.

The variations in the amplification ratios at the slope crest were complicated at varying incident angles. The horizontal amplification ratios at the left crest decreased with an increase in the angle of incidence, and the reduced amplitude intensified as the slope became steeper ($i < 1.0$). The horizontal amplification ratios at the right crest remained the same with an increase in the angle of incidence, and the maximum ratios fell between 1.0

and 1.5 of *i*. The vertical amplification ratios at the left crest and right crest decreased with an increase in slope inclination. The ratios were intensified at the left crest whereas they were alleviated at a slope inclination of 1.0, at an incident angle of 15°on the right crest.

In summary, the variations in the amplification ratios were concentrated on the slope crests at varying angles of incidence because the varying inclinations of slope ridges changed the propagation of wave paths in the topography.

5. Summary and Conclusions

In this study, ground motion amplification of the slope topography was analyzed using dynamic FEMs. The wave patterns, materials and slope sizes were discussed based on varying angles of incidence. A viscous-spring artificial boundary was borrowed, and an equivalent nodal force method was proposed to implement oblique incident waves in the FEM. Two numerical examples were adopted to verify the validity of the input method and the accuracy of the artificial boundary. Subsequently, the ground motions in slope topography were investigated with arbitrary directions of incidence, considering the main impact factors of site effects (wave patterns, materials and sizes of slope topography).

The main conclusions and findings are as follows: (1) the amplification effects are underestimated by the vertical incident waves, and the maximum ground motions are closely related to the scattered waves, which depend on the ridge inclinations and the incident directions. (2) Owing to the intense amplification effects under SV waves, the ground motions of slope topography are preferred for analysis using SV waves rather than P waves. (3) With an increase in the angles of incidence, the amplification effects are more complicated at the crest of the slope topography, and the amplification regions are focused on the ridges away from the epicenter. (4) The ground motion amplification effects are aggravated in soft materials owing to the much more scattered waves produced. The directions of the incident waves are a key factor in the analysis of ground motions in slope topography. The amplification effects of slope topography should be discussed separately owing to the complex transmission paths induced by oblique incident waves.

The results obtained here can be considered in the analysis of the ground motion amplification effects of slope topography subjected to earthquakes. The above findings were discussed in terms of three influencing factors that impact the topographic effects: the patterns of the incident waves, the materials of the slope and the slope sizes. The wave patterns that included P waves and SV waves were sufficient for the analysis of topographic effects. Even though the surface waves (e.g., Rayleigh and Love waves) have not been directly analyzed, they were still considered in this study because the conclusions were based on the incident body waves, that is, P waves or SV waves, and were actually a comprehensive result of body waves and surface waves. That is, the proposed input method can also express the impacts of Rayleigh waves (in two dimensions). Thus, the topographic effects calculated using P waves and SV waves in this study are meaningful and necessary. Even though the samples in the analysis of slope materials and slope sizes were relatively few, the conclusions could interpret several regularity phenomena, achieving the purpose of this study. Furthermore, the lack of experimental evidence for the comparison and calibration of the findings in the numerical studies is a serious obstacle [36,61]. Hence, a specific model should be established according to the study area while considering real materials based on the specific earthquake record for a specific study area. Data from field measurements should be collected simultaneously to support numerical modeling results.

Author Contributions: Conceptualization, C.Y.; methodology, C.Y. and W.-H.L.; software, C.Y.; validation, C.Y. and W.W.; formal analysis, C.Y. and W.-H.L.; writing—original draft preparation, C.Y. writing—review and editing, C.Y.; funding acquisition, C.Y. All authors have read and agreed to the published version of the manuscript.

Funding: This research was supported by the Postdoctoral Science Foundation of China (2021M692242); the National Natural Science Foundation of Hebei Province (E2021210072, E2021210036); and the Central Leading Local Science and Technology Development Foundation of Hebei Province (216Z5403G). The support is gratefully acknowledged.

Institutional Review Board Statement: Not applicable.

Informed Consent Statement: Not applicable.

Data Availability Statement: Not applicable.

Conflicts of Interest: The authors declare no conflict of interest.

References

1. Alfaro, P.; Delgado, J.; García-Tortosa, F.J.; Lenti, L.; López, J.A.; López-Casado, C.; Martino, S. Widespread landslides induced by the Mw 5.1 earthquake of 11 May 2011 in Lorca, SE Spain. *Eng. Geol.* **2012**, *137-138*, 40–52. [CrossRef]
2. Collins, B.D.; Kayen, R.; Tanaka, Y. Spatial distribution of landslides triggered from the 2007 Niigata Chuetsu–Oki Japan Earthquake. *Eng. Geol.* **2012**, *127*, 14–26. [CrossRef]
3. Delgado, J.; García-Tortosa, F.J.; Garrido, J.; Loffredo, A.; López-Casado, C.; Martin-Rojas, I.; Rodríguez-Peces, M.J. Seismically-induced landslides by a low-magnitude earthquake: The M$_w$ 4.7 Ossa De Montiel event (central Spain). *Eng. Geol.* **2015**, *196*, 280–285. [CrossRef]
4. Tang, C.; Ma, G.C.; Chang, M.; Li, W.L.; Zhang, D.D.; Jia, T.; Zhou, Z.Y. Landslides triggered by the 20 April 2013 Lushan earthquake, Sichuan Province, China. *Eng. Geol.* **2015**, *187*, 45–55. [CrossRef]
5. Wartman, J.; Dunham, L.; Tiwari, B.; Pradel, D. Landslides in eastern Honshu induced by the 2011 Tohoku earthquake. *Bull. Seism. Soc. Am.* **2013**, *103*, 1503–1521. [CrossRef]
6. Xu, C.; Xu, X.W.; Shyu, J.B.H. Database and spatial distribution of landslides triggered by the Lushan, China Mw 6.6 earthquake of 20 April 2013. *Geomorphology* **2015**, *248*, 77–92. [CrossRef]
7. Dunning, S.A.; Mitchell, W.A.; Rosser, N.J.; Petley, D.N. The Hattian Bala rock avalanche and associated landslides triggered by the Kashmir Earthquake of 8 October 2005. *Eng. Geol.* **2007**, *93*, 130–144. [CrossRef]
8. Owen, L.A.; Kamp, U.; Khattak, G.A.; Harp, E.L.; Keefer, D.K.; Bauer, M.A. Landslides triggered by the 8 October 2005 Kashmir earthquake. *Geomorphology* **2008**, *94*, 1–9. [CrossRef]
9. Yin, Y.P.; Wang, F.W.; Sun, P. Landslide hazards triggered by the 2008 Wenchuan earthquake, Sichuan, China. *Landslides* **2009**, *6*, 139–152. [CrossRef]
10. Huang, R.; Pei, X.; Fan, X.; Zhang, W.; Li, S.; Li, B. The characteristics and failure mechanism of the largest landslide triggered by the Wenchuan earthquake, May 12, 2008, China. *Landslides* **2011**, *9*, 131–142. [CrossRef]
11. Sepúlveda, S.A.; Murphy, W.; Jibson, R.W.; Petley, D.N. Seismically induced rock slope failures resulting from topographic amplification of strong ground motions: The case of Pacoima Canyon, California. *Eng. Geol.* **2005**, *80*, 336–348. [CrossRef]
12. Sepúlveda, S.A.; Murphy, W.; Petley, D.N. Topographic controls on coseismic rock slides during the 1999 Chi-Chi earthquake, Taiwan. *Q. J. Eng. Geol. Hydrogeol.* **2005**, *38*, 189–196. [CrossRef]
13. Bourdeau, C.; Havenith, H.B. Site effects modelling applied to the slope affected by the Suusamyr earthquake (Kyrgyzstan, 1992). *Eng. Geol.* **2008**, *97*, 126–145. [CrossRef]
14. Bozzano, F.; Lenti, L.; Martino, S.; Paciello, A.; Mugnozza, G.S. Self-excitation process due to local seismic amplification responsible for the reactivation of the Salcito landslide (Italy) on 31 October 2002. *J. Geophys. Res. Space Phys.* **2008**, *113*, 1–21. [CrossRef]
15. Del Gaudio, V.; Coccia, S.; Wasowski, J.; Gallipoli, M.R.; Mucciarelli, M. Detection of directivity in seismic site response from microtremor spectral analysis. *Nat. Hazards Earth Syst. Sci.* **2008**, *8*, 751–762. [CrossRef]
16. Bozzano, F.; Lenti, L.; Martino, S.; Paciello, A.; Mugnozza, G.S. Evidences of landslide earthquake triggering due to self-excitation process. *Int. J. Earth. Sci.* **2011**, *100*, 861–879. [CrossRef]
17. Lenti, L.; Martino, S. The interaction of seismic waves with step-like slopes and its influence on landslide movements. *Eng. Geol.* **2012**, *126*, 19–36. [CrossRef]
18. Athanasopoulos, G.A.; Pelekis, P.C.; Leonidou, E.A. Effects of surface topography on seismic ground response in the Egion (Greece) 15 June 1995 earthquake. *Soil. Dyn. Earthq. Eng.* **1999**, *18*, 135–149. [CrossRef]
19. Çelebi, M. Topographical and geological amplification: Case studies and engineering implications. *Struct. Saf.* **1991**, *10*, 199–217. [CrossRef]
20. Wang, W.; Liu, B.D.; Liu, X.; Yang, M.L.; Zhou, Z.H. Analysis on the hill topography effect based on the strong ground motion records of Wenchuan Ms8.0 earthquake. *Acta. Seismol. Sinica.* **2015**, *37*, 452–462.
21. Zaslavsky, Y.; Shapira, A. Experimental study of topographic amplification using the Israel seismic network. *J. Earthq. Eng.* **2000**, *4*, 43–65. [CrossRef]
22. Bakavoli, M.K.; Haghshenas, E. Experimental and numerical study of topographic site effect on a hill near Tehran. In Proceedings of 5th International Conference on Recent Advances in Geotechnical Earthquake Engineering and Soil Dynamics, San Diego, CA, USA, 24 May 2010; pp. 1–10.
23. Trifunac, M.D. Scattering of plane SH waves by a semi-cylindrical canyon. *Earthq. Eng. Struct. Dyn.* **1973**, *1*, 267–281. [CrossRef]

24. Wong, H.L.; Trifunac, M.D. Scattering of plane SH waves by a semi-elliptical canyon. *Earthq. Eng. Struct. Dyn.* **1974**, *3*, 157–169. [CrossRef]
25. Lee, V.W. Three-dimensional diffraction of plane P, SV & SH waves by a hemispherical alluvial valley. *Soil Dyn. Earthq. Eng.* **1984**, *3*, 133–144. [CrossRef]
26. Lee, V.W. Scattering of plane SH-waves by a semi-parabolic cylindrical canyon in an elastic half-space. *Geophys. J. Int.* **1990**, *100*, 79–86. [CrossRef]
27. Yuan, X.M.; Men, F.L. Scattering of plane SH waves by a semi-cylindrical hill. *Earthq. Eng. Struct. Dyn.* **1992**, *21*, 1091–1098. [CrossRef]
28. Du, X.L.; Xiong, J.G.; Guan, H.M. Boundary integration equation method to scattering of plane SH-waves. *Acta Seism. Sin.* **1993**, *6*, 609–618. [CrossRef]
29. Ashford, S.A.; Sitar, N. Analysis of topographic amplification of inclined shear waves in a steep coastal bluff. *Bull. Seismol. Soc. Am.* **1997**, *87*, 692–700. [CrossRef]
30. Ashford, S.A.; Sitar, N.; Lysmer, J.; Deng, N. Topographic effects on the seismic response of steep slopes. *Bull. Seism. Soc. Am.* **1997**, *87*, 701–709.
31. Han, F.; Wang, G.Z.; Kang, C.Y. Scattering of SH-waves on triangular hill joined by semi-cylindrical canyon. *Appl. Math. Mech.* **2011**, *32*, 309–326. [CrossRef]
32. Yang, Z.L.; Xu, H.N. Ground motion of two scalene triangle hills and a semi-cylindrical canyon under incident SH-waves. *Appl. Mech. Mater.* **2012**, *21–126*, 862–866. [CrossRef]
33. Lee, V.W.; Liu, W.Y. Two-dimensional scattering and diffraction of P- and SV-waves around a semi-circular canyon in an elastic half-space: An analytic solution via a stress-free wave function. *Soil Dyn. Earthq. Eng.* **2014**, *63*, 110–119. [CrossRef]
34. Zhang, N.; Gao, Y.F.; Pak, R.Y.S. Soil and topographic effects on ground motion of a surficially inhomogeneous semi-cylindrical canyon under oblique incident SH waves. *Soil. Dyn. Earthq. Eng.* **2017**, *95*, 17–28. [CrossRef]
35. Meunier, P.; Hovius, N.; Haines, J.A. Topographic site effects and the location of earthquake induced landslides. *Earth Planet. Sci. Lett.* **2008**, *275*, 221–232. [CrossRef]
36. Bouckovalas, G.D.; Papadimitriou, A.G. Numerical evaluation of slope topography effects on seismic ground motion. *Soil Dyn. Earthq. Eng.* **2005**, *25*, 547–558. [CrossRef]
37. Pagliaroli, A.; Lanzo, G.; Beniamino, E. Numerical evaluation of topographic effects at the nicastro ridge in Southern Italy. *J. Earthq. Eng.* **2011**, *15*, 404–432. [CrossRef]
38. Narayan, J.P.; Kumar, V. Study of combined effects of sediment rheology and basement focusing in an unbounded viscoelastic medium using P-SV-Wave finite-difference modeling. *Acta Geophys.* **2014**, *62*, 1214–1245. [CrossRef]
39. Hailemikael, S.; Lenti, L.; Martino, S.; Paciello, A.; Rossi, D.; Mugnozza, G.S. Ground-motion amplification at the Colle di Roio ridge, Central Italy: A combined effect of stratigraphy and topography. *Geophys. J. Int.* **2016**, *206*, 1–18. [CrossRef]
40. Assimaki, D.; Gazetas, G.; Kausel, E. Effects of local soil conditions on the topographic aggravation of seismic motion: Parametric investigation and recorded field evidence from the 1999 Athens earthquake. *Bull. Seism. Soc. Am.* **2005**, *95*, 1059–1089. [CrossRef]
41. Assimaki, D.; Kausel, E.; Gazetas, G. Wave propagation and soil structure interaction on a cliff crest during the 1999 Athens earthquake. *Soil. Dyn. Earthq. Eng.* **2005**, *25*, 513–527. [CrossRef]
42. Lo Presti, D.C.F.; Lai, C.; Puci, I. ONDA: Computer code for nonlinear seismic response analyses of soil deposits. *J. Geotech. Geoenviron. Eng.* **2006**, *132*, 223–235. [CrossRef]
43. Di Fiore, V. Seismic site amplification induced by topographic irregularity: Results of a numerical analysis on 2D synthetic models. *Eng. Geol.* **2010**, *114*, 109–115. [CrossRef]
44. Tripe, R.; Kontoe, S.; Wong, T.K.C. Slope topography effects on ground motion in the presence of deep soil layers. *Soil Dyn. Earthq. Eng.* **2013**, *50*, 72–84. [CrossRef]
45. Rizzitano, S.; Cascone, E.; Biondi, G. Coupling of topographic and stratigraphic effects on seismic response of slopes through 2D linear and equivalent linear analyses. *Soil Dyn. Earthq. Eng.* **2014**, *67*, 66–84. [CrossRef]
46. Nguyen, K.V.; Gatmiri, B. Evaluation of seismic ground motion induced by topographic irregularity. *Soil Dyn. Earthq. Eng.* **2007**, *27*, 183–188. [CrossRef]
47. Pagliaroli, A.; Avalle, A.; Falcucci, E.; Gori, S.; Galadini, F. Numerical and experimental evaluation of site effects at ridges characterized by complex geological setting. *Bull. Earthq. Eng.* **2015**, *13*, 2841–2865. [CrossRef]
48. Gischig, V.S.; Eberhardt, E.; Moore, J.R.; Hungr, O. On the seismic response of deep-seated rock slope instabilities—Insights from numerical modeling. *Eng. Geol.* **2015**, *193*, 1–18. [CrossRef]
49. Narayan, J.P.; Kumar, V. A numerical study of effects of ridge-weathering and ridge-shape-ratio on the ground motion characteristics. *J. Seism.* **2015**, *19*, 83–104. [CrossRef]
50. Li, H.; Liu, Y.; Liu, L.; Liu, B.; Xia, X. Numerical evaluation of topographic effects on seismic response of single-faced rock slopes. *Bull. Int. Assoc. Eng. Geol. Environ.* **2017**, 1873–1891. [CrossRef]
51. Jin, X.; Liao, Z.P. Statistical research on S-wave incident angle. *Earthq. Resear. Chin.* **1994**, *8*, 121–131.
52. Takahiro, S.; Kazuhiko, K.; Yoichi, S.; Dai, K.; Toru, M.; Ryoichi, T.; Chiaki, Y.; Takeshi, U. Estimation of earthquake motion incident angle at rock site. In Proceedings of the 12th World Conference Earthquake Engineering, NZ National Society for Earthquake Engineering, Auckland, New Zealand, 16 April 2000; pp. 1–8.

53. Kuhlemeyer, R.L.; Lysmer, J. Finite element method accuracy for wave propagation problems. *J. Soil Mech. Found. Div.* **1973**, *99*, 421–427. [CrossRef]
54. Deeks, A.J.; Randolph, M.F. Axisymmetric time-domain transmitting boundaries. *J. Eng. Mech.* **1994**, *120*, 25–42. [CrossRef]
55. Liu, J.B.; Lv, Y.D. A direct method for analysis of dynamic soil-structure interaction based on interface idea. *Chin. Civ. Eng. J.* **1998**, *83*, 261–276. [CrossRef]
56. Du, X.L.; Zhao, M.; Wang, J.T. A stress artificial boundary in FEA for near-field wave problem. *Chin. J. Theor. Appl. Mech.* **2006**, *38*, 49–56.
57. Yin, C.; Li, W.H.; Zhao, C.G.; Kong, X.A. Impact of tensile strength and incident angles on a soil slope under earthquake SV-waves. *Eng. Geol.* **2019**, *260*, 1–11. [CrossRef]
58. Eringen, A.C.; Suhubi, E.S. *Elastodynamics*; Academic Press: New York, NY, USA, 1975.
59. Ewing, W.M.; Jardetzky, W.S.; Press, F.; Beiser, A. *Elastic Waves in Layered Media*; McGraw-Hill: New York, NY, USA, 1957.
60. Wong, H.L. Effect of surface topography on the diffraction of P, SV, and Rayleigh waves. *Bull. Seismol. Soc. Am.* **1982**, *72*, 1167–1183.
61. Lenti, L.; Martino, S. A parametric numerical study of the interaction between seismic waves and *Landslides* for the evaluation of the susceptibility to seismically induced displacements. *Bull. Seismol. Soc. Am.* **2013**, *103*, 33–56. [CrossRef]

Article

Microcrack Porosity Estimation Based on Rock Physics Templates: A Case Study in Sichuan Basin, China

Chuantong Ruan [1,2], Jing Ba [1,*], José M. Carcione [1,3], Tiansheng Chen [4] and Runfa He [1]

[1]   School of Earth Sciences and Engineering, Hohai University, Nanjing 211100, China; rct@hhu.edu.cn (C.R.); jcarcione@libero.it (J.M.C.); herunfa@hhu.edu.cn (R.H.)

[2]   School of Mathematics and Statistics, Zhoukou Normal University, Zhoukou 466001, China

[3]   National Institute of Oceanography and Applied Geophysics (OGS), 34010 Trieste, Italy

[4]   Petroleum Exploration and Production Research Institute, SINOPEC, Beijing 100083, China; chents.syky@sinopec.com

*   Correspondence: jba@hhu.edu.cn

Abstract: Low porosity-permeability structures and microcracks, where gas is produced, are the main characteristics of tight sandstone gas reservoirs in the Sichuan Basin, China. In this work, an analysis of amplitude variation with offset (AVO) is performed. Based on the experimental and log data, sensitivity analysis is performed to sort out the rock physics attributes sensitive to microcrack and total porosities. The Biot–Rayleigh poroelasticity theory describes the complexity of the rock and yields the seismic properties, such as Poisson's ratio and P-wave impedance, which are used to build rock-physics templates calibrated with ultrasonic data at varying effective pressures. The templates are then applied to seismic data of the Xujiahe formation to estimate the total and microcrack porosities, indicating that the results are consistent with actual gas production reports.

Keywords: tight sandstone gas reservoirs; rock-physics template; Biot–Rayleigh theory; total porosity; microcrack porosity

Citation: Ruan, C.; Ba, J.; Carcione, J.M.; Chen, T.; He, R. Microcrack Porosity Estimation Based on Rock Physics Templates: A Case Study in Sichuan Basin, China. *Energies* **2021**, *14*, 7225. https://doi.org/10.3390/en14217225

Academic Editors: Joel Sarout and Manoj Khandelwal

Received: 23 August 2021
Accepted: 27 October 2021
Published: 2 November 2021

Publisher's Note: MDPI stays neutral with regard to jurisdictional claims in published maps and institutional affiliations.

1. Introduction

The development of society has led to a significant increase in the demand of oil and gas resources, and the depletion of conventional petroleum resources made the exploration and extraction of unconventional hydrocarbon resources necessary [1,2]. Tight sandstone reservoirs are widely distributed and account for a high production of China's total natural gas [3,4]. Large-scale gas fields with great potential have been discovered in more than ten basins, including Sichuan, Ordos, Songliao, Tuha, and Junggar [5].

Deep burial generally leads to high mechanical compaction and cementation of a sandstone reservoir, so that tight and heterogeneous reservoir rocks are generated, where microcracks are developed [6]. These microcracks affect the rock elastic properties and control the fluid flow, providing channels for hydrocarbon migration [7–9]. Their identification is a key factor for evaluating reservoir quality [10,11].

The study of microcracks in tight sandstones has become an important topic of rock-physics studies [12]. Hudson [13] established a relation between microcrack density and rock elastic properties, while Smith [14] stated that in low porosity rocks, the effect of cracks on seismic velocity can be important as mineral composition, total porosity, and fluid type. Yoon [15] established a relation between aspect ratio and microcrack density. Cheng [16] studied the effect of effective pressure and fluids. Zhang et al. [17,18] proposed a differential poroelastic model to describe wave propagation and dissipation in fluid-saturated rocks which contain inclusions at multiple scales.

The use of RPT (rock physics template; the list of abbreviations is given in Abbreviations) is widespread: Avseth and Ødegaard [19] linked the reservoir characteristics with elastic attributes, whereas Avseth et al. [20] used the Biot–Gassmann theory to discriminate

85

between oil and gas in reservoirs. RPT was also used by Xin and Han [21] to estimate reservoir lithology and fluids. Michel [22] predicted porosity and oil saturation in shales, based on RPTs built with P-wave impedance (I_P) and the ratio between the P- and S-wave velocities (V_P/V_S). Gupta et al. [23] identified thin sandy layers saturated with oil, and Ba et al. [24] estimated porosity and gas saturation. Carcione and Avseth [25] evaluated organic content, hydrocarbon saturation, and in-situ pressure in source rocks, and Liu et al. [26] proposed a combination of fluid indicators. Pang et al. [27] predicted microcrack properties in tight sandstone reservoirs, based on P-wave impedance and attenuation.

The Biot–Rayleigh (BR) theory was introduced by Ba et al. [28,29] for describing wave propagation characteristics in rocks with multi-phase pore structures. This work is mainly based on BR theory. We consider ten wells in the proposed area and perform a sensitivity analysis by using the ultrasonic and well-log data. On the basis of Poisson's ratio (ν) and P-wave impedance, we establish multi-scale RPTs for tight sandstone gas reservoirs, which relate the total and microcrack porosities to the seismic properties, which are obtained by means of seismic inversion.

2. Overview of the Work Area

2.1. Geology

The West Sichuan depression, bounded by the Longmenshan thrust and Longquanshan uplift belts, is a key area of natural gas exploration in China [30]. The maximum buried depth of the Xujiahe Formation exceeds 4.5 km, and the average thickness of the target layer is approximately 120 m. The work area was subjected to multi-stage Indosinian, Yanshan, and Himalayan tectonic movements, which resulted in diverse structural fractures that are controlled by complex diagenesis processes. A mechanical compaction is an aspect of diagenesis that destroys the primary pores. The sedimentary facies are mainly about the braided river delta facies. The channels overlap with each other from different periods, forming a longitudinal superposition and plane contiguous sand bodies [31].

2.2. Reservoir Characteristics

The tight sandstones are mainly composed of litharenites, with small amounts of lithic quartz arenite and feldspathic litharenite. In terms of mineral content, the average amounts of quartz and feldspar are 69% and 8.2%, respectively [32]. The effect of compaction on the reservoirs is more significant than that of cementation. Figure 1 shows that the compaction process is responsible for the developed geological characteristics such as complex pore structures and network microcracks. The average secondary, primary, and microcrack porosities are 2.9%, 1.1%, and 0.3%, respectively. The grain size varies from fine to medium, and the sorting property varies from poor to good. Moreover, the reservoir exhibits low porosity, low matrix permeability, and small pore-throat radii [31]. The microcracks maintain the reservoir connectivity and facilitate the gas production [33].

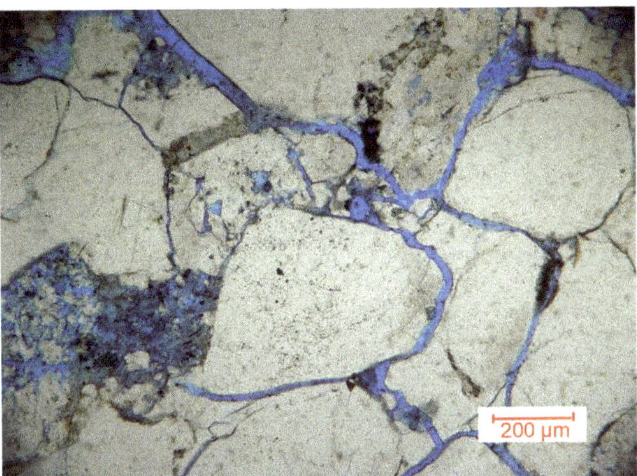

Figure 1. Thin section of a tight sandstone showing microcracks and a complex pore structure.

2.3. AVO Characteristics

Amplitude variations with offset (AVO) discriminate lithologies and help to detect hydrocarbons [34]. Rutherford and Williams [35] categorized the AVO responses of mudstone/gas-bearing sandstone interface into the three types. Subsequently, Castagna and Swan [36] added a fourth AVO response type. We analyze the AVO characteristics based on the profiles of Well P, shown in Figure 2, where the target layer is indicated with dashed red lines. Table 1 shows the seismic properties of the sandstone model of the Xujiahe Formation based on the log data. To obtain the P- and S-wave velocities and density, a 60 m interval is selected. The model data are collected from the interval above the top interface of the target layer that has a relatively stable P-wave velocity. For the target layer, we consider average values.

Figure 3 illustrates the AVO of the top (red color) and bottom (blue color) interfaces. The characteristics of the curves agree with the fact that that the P- and S-wave impedances of the target layer are smaller than those of the overburden mudstone. The top and bottom responses are type IV and I AVO curves, respectively.

Table 1. Seismic properties, corresponding to the sandstone model of the Xujiahe Formation.

Well	Top			Target Layer			Bottom		
	V_P (m/s)	V_S (m/s)	Density (g/cm^3)	V_P (m/s)	V_S (m/s)	Density (g/cm^3)	V_P (m/s)	V_S (m/s)	Density (g/cm^3)
P	4453	2812	2.5360	3956	2462	2.5298	4787	2983	2.5531

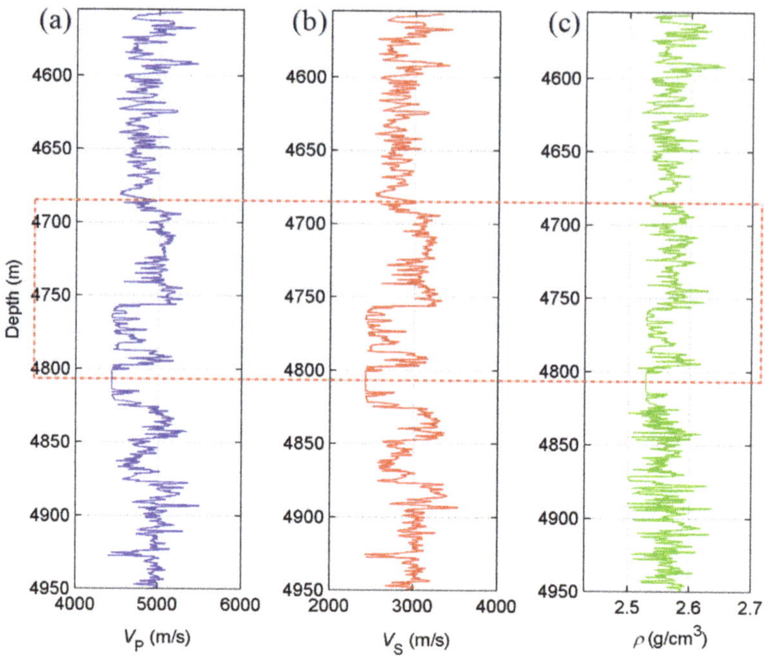

Figure 2. Log profiles of Well P. (**a**) P-wave velocity, (**b**) S-wave velocity, and (**c**) density.

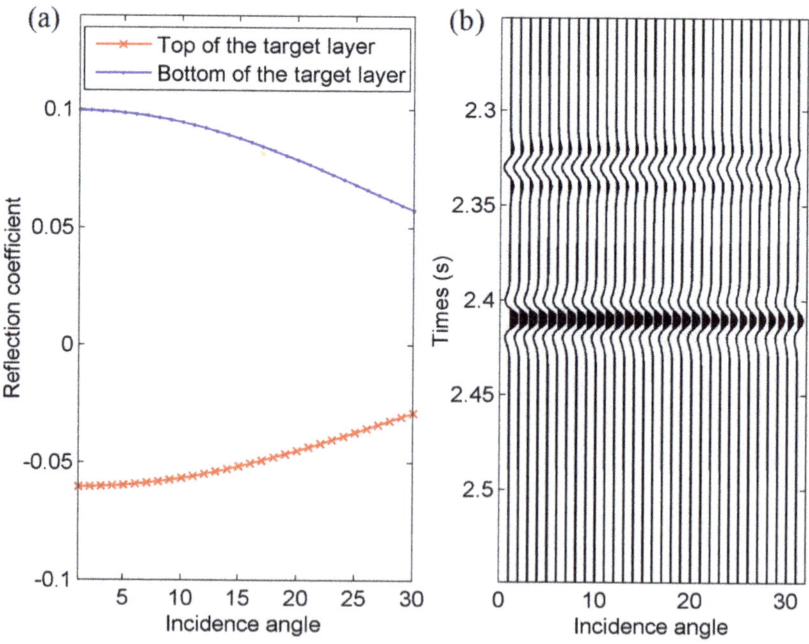

Figure 3. (**a**) AVO curves and (**b**) synthetic seismic records of the target layer.

3. Theory and Flowchart

Figure 4 shows an idealization of the tight sandstone of the Xujiahe Formation, which is characterized by a complex pore structure and microcrack network. The total porosity is the sum of intergranular and microcrack porosity.

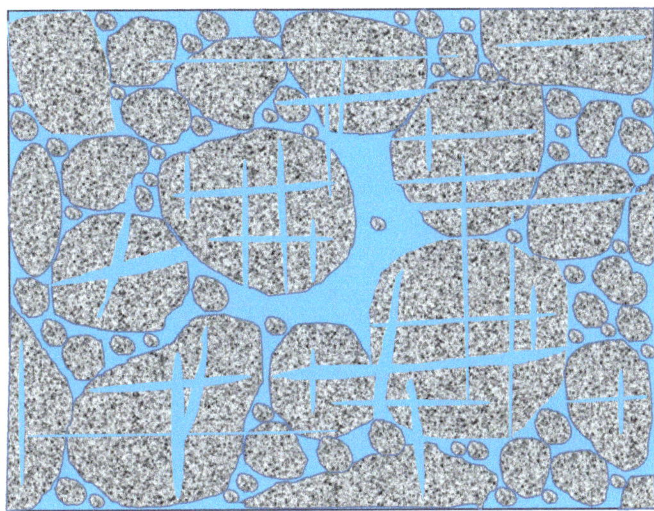

Figure 4. Scheme of tight sandstone pore structure with intergranular pores and microcrack network.

The flowchart for the prediction of the reservoir properties is given in Figure 5. The microcrack porosity is estimated by using the quantitative relation between the elastic attributes and reservoir properties, based on a rock-physics model and aided by geological, log data, and ultrasonic data.

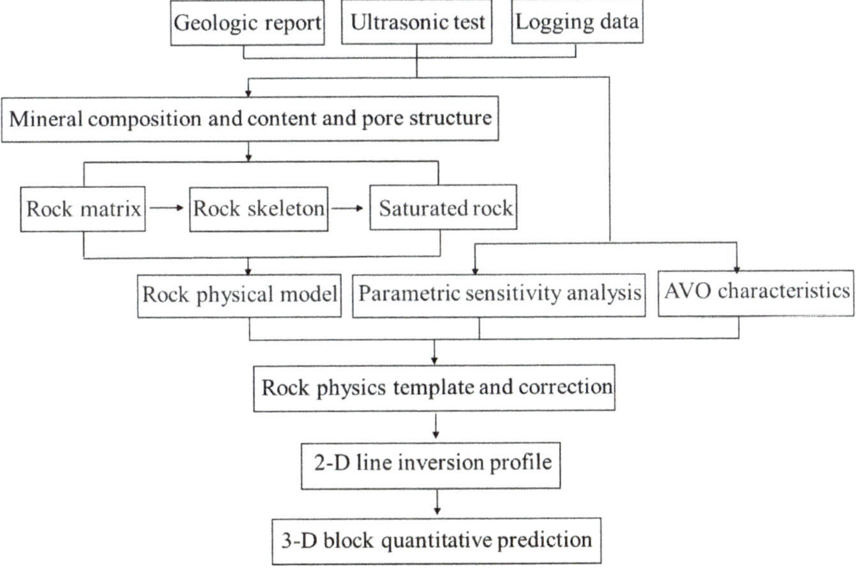

Figure 5. Flowchart for the prediction of the reservoir properties.

The main minerals are quartz, feldspar, and clay. The total porosity is less than 12%. The procedure to establish the rock-physics model is as follows.

(1) The Voigt–Reuss–Hill equation [37–39] is used to compute the elastic modulus of mineral mixture M_{VRH} according to the mineral composition:

$$M_V = \sum_{i=1}^{N} f_i M_i,$$

(1)

$$\frac{1}{M_R} = \sum_{i=1}^{N} f_i / M_i,$$

(2)

$$M_{VRH} = \frac{M_V + M_R}{2},$$

(3)

where f_i and M_i denote the volume fraction and elastic modulus, respectively, of the i-th component.

(2) According to the pore structure shown in Figure 4, we use the differential equivalent medium (DEM) theory [40] to add spherical pores and oblate cracks, whose aspect ratio are 1 and 0.0005, respectively, to add pores and microcracks into the matrix, and obtain the bulk and shear moduli of the rock skeleton (starred quantities),

$$(1 - y)d/dy[K^*(y)] = (K_2 - K^*(y))P^{(*2)}(y),$$

(4)

$$(1 - y)d/dy[\mu^*(y)] = (\mu_2 - \mu^*(y))Q^{(*2)}(y).$$

(5)

The initial conditions are $K^*(0) = K_1$ and $\mu^*(0) = \mu_1$, where K_1 and μ_1 are the bulk and shear moduli, respectively, of the initial main material (phase 1); K_2 and μ_2 are the bulk and shear moduli, respectively, of the inclusion that is gradually added (phase 2); y is the volume content of phase 2; and $P^{(*2)}$ and $Q^{(*2)}$ [41] are related to the shape of the inclusion.

(3) The equations proposed by Batzle and Wang [42] are used to compute the bulk modulus, viscosity, and density of pore fluids at different temperatures and pressures.

(4) The BR theory is used to compute the wave response. Appendix A shows the computation of the complex wave number k. The complex velocity is $v = \omega/k$, where ω is the angular frequency, and the P-wave velocity and quality factor are [7]

$$V_P = \left[\text{Re}\left(v^{-1}\right)\right]^{-1},$$

(6)

$$Q = \frac{\text{Re}\left(v^2\right)}{\text{Im}\left(v^2\right)},$$

(7)

where "Re" and "Im" take real and imaginary parts, respectively.

We consider the inclusion radius of 100 μm, and the bulk and shear moduli of the matrix are 33 GPa and 45 GPa, respectively. Gas has a bulk modulus of 0.02 GPa, a density of 0.089 g/cm^3, and viscosity of 0.016×10^{-3} Pa·s. The crack porosity is set to 0.2%, and the total porosity varies as shown in Figure 6, where the P-wave velocity and dissipation factor are plotted as a function of frequency. Increasing porosity implies increasing velocity dispersion, attenuation, and relaxation frequency.

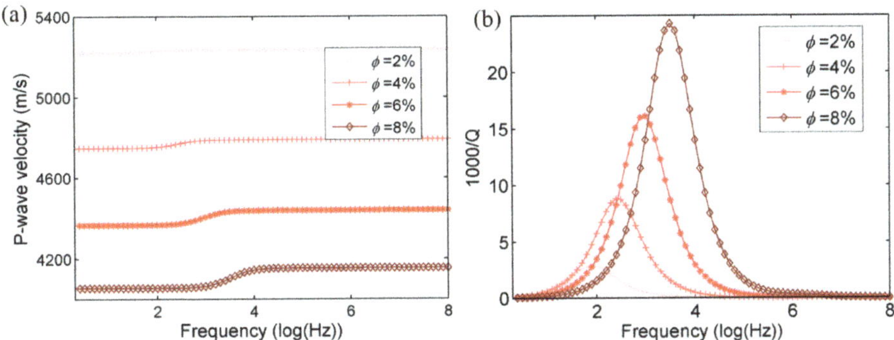

Figure 6. P-wave velocity (**a**) and dissipation factor (**b**) as a function of frequency and different total porosities.

Figure 7 shows the P- and S-wave velocities as a function of the total porosity for various microcrack porosities at 1 MHz. As can be seen, the velocities decrease with increasing total and microcrack porosities, as expected.

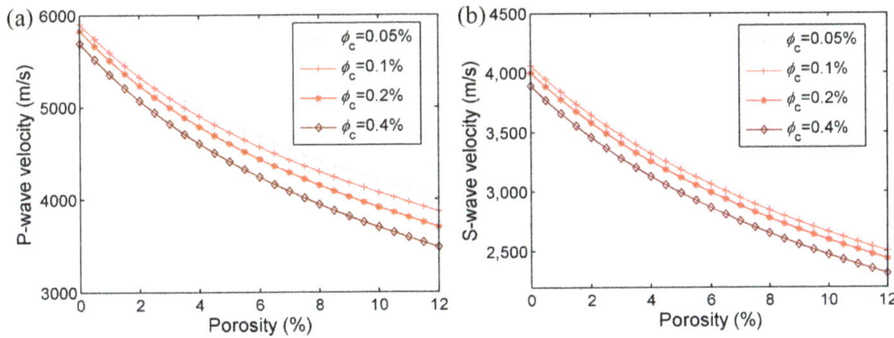

Figure 7. Effect of the total and microcrack porosities on the P-wave (**a**) and S-wave (**b**) velocities.

4. Ultrasonic Experiments and Sensitivity Analysis

To investigate the effects of the microcracks, ultrasonic experiments at 1 MHz were performed. We select a sample with a porosity of 4.39% and a grain bulk modulus of 39 MPa. The grain, dry-rock, and wet-rock densities are 2.691, 2.573, and 2.62 g/cm^3, respectively, where the sample is saturated with water. The experimental setup proposed by Guo et al. [43] is used to measure the velocities at 20 °C, with the ultrasonic pulse method. The sample is sealed with a rubber sleeve and placed in the vessel. The pore pressure and temperature are fixed. Effective pressures of 5, 10, 15, 20, 25, 30, and 35 MPa are applied to the sample, in both the gas-saturated and water-saturated cases, and the velocities are measured.

As shown in Figure 8a, the measured P- and S-wave velocities increase with effective pressure. The increase of the confining pressure at a fixed pore pressure will lead to the gradual closure of internal microcracks (especially at low effective pressures), so as to stiffen the rock skeleton and increase its elastic moduli (and the wave velocities).

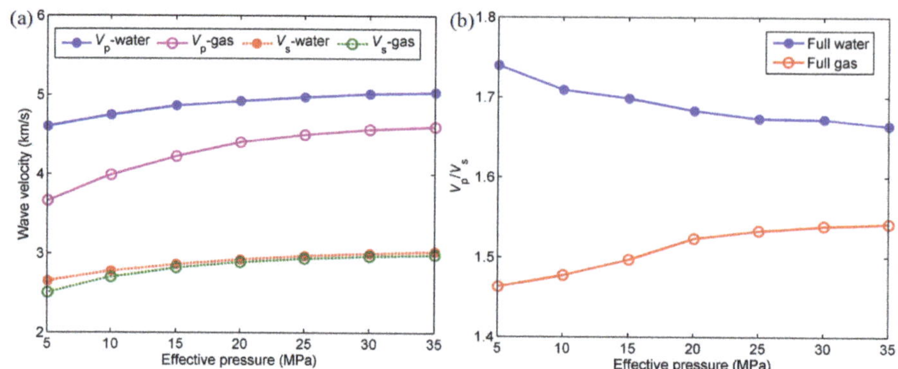

Figure 8. P- and S-wave velocities (**a**) and their ratio (**b**) as a function of effective pressure.

Figure 8b shows the velocity ratio. In the full gas saturation case, V_P/V_S increases with increasing effective pressure and the opposite behavior occurs for full water saturation case.

In addition, we measured the total porosity in the range 5–35 MPa. Microcracks with small aspect ratios tend to close first with increasing effective pressure and the relation between porosity and pressure changes from exponential to linear [44].

Experimental measurements can be performed to predict microcrack porosity based on the relation between porosity and effective pressure [45–47]. Stiff porosity can be obtained through a linear extrapolation of this relation and the microcrack porosity can be estimated as the difference between the total and stiff porosities. Figure 9 shows the different porosities as a function of the effective pressure.

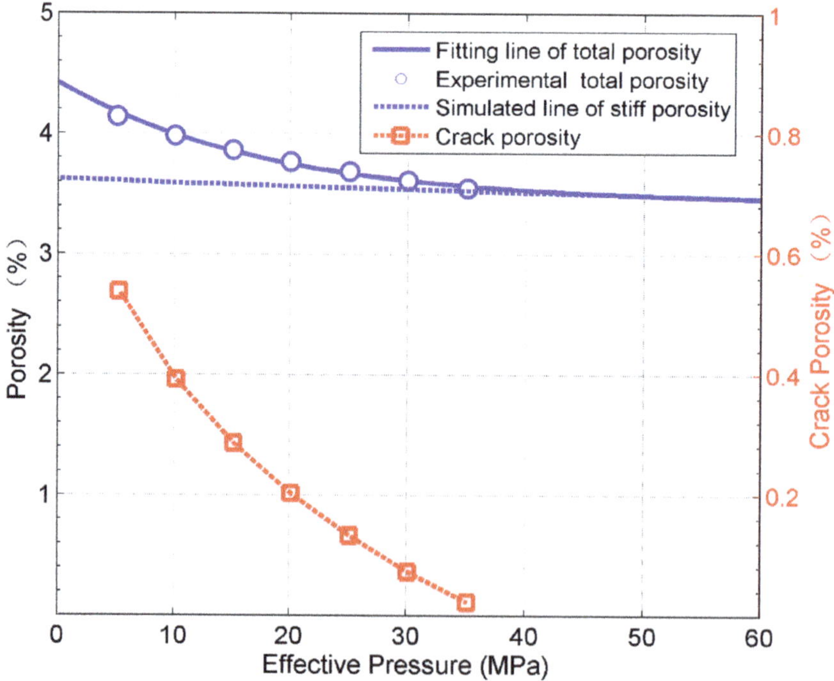

Figure 9. Total, stiff, and microcrack porosities as a function of the effective pressure.

The sensitivity analysis method like Sobol's indices can be used to analyze the interactions between the parameters, which can be applied with the given forward modeling equation or method/procedure [48]. In this work, sensitivity analysis to the total and microcrack porosities as inputs is performed based on the actually measured data from the experiments and well logs as puts, without considering the detailed modeling equations (or procedures). Based on the observed experimental or log data, the fluid sensitivity indicator (FSI) [49–52] has been proposed to analyze how the rock elastic properties are affected by the different fluid saturation statuses. Whereas, in this study, a similar method is adopted to analyze the relative variations of rock physics attributes with respect to the changes in total/microcrack porosity, and it is defined as

$$SI = \frac{\left|\overline{A} - A_m\right|}{A_m},\tag{8}$$

where A_m and $\overline{A}$ represent the value of rock physics attribute at the minimal total (or microcrack) porosity and the average value of rock physics attribute within the considered range, respectively. Eleven attributes are considered, including density and the P- and S-wave velocities. The other eight are $V_\mathrm{P}/V_\mathrm{S}$, P-wave impedance ($I_\mathrm{P}$), S-wave impedance ($I_\mathrm{S}$), Poisson's ratio ($\nu$), shear modulus ($\mu$), the first Lamé constant (λ), $\lambda\rho$, and V_P/ρ.

Figure 10 shows the sensitivity indices to microcrack porosity at ultrasonic frequencies, where we observe that $\lambda\rho$, λ, ν, μ, and I_P are the most sensitive ones.

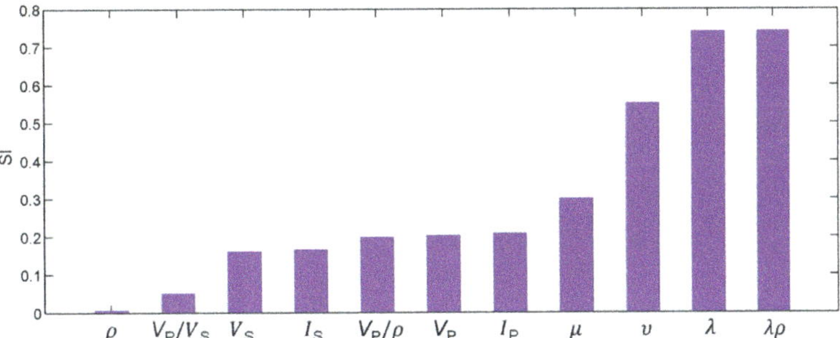

Figure 10. Sensitivity indices varying with the microcrack porosity.

The sensitivity analysis to total porosity is performed for the sonic log data, where we consider the gas-saturated layers of Well P. Figure 11 shows the results, where $\lambda\rho$, λ, ν, μ, and I_P are the most sensitive. In Figure 10, V_P/ρ, V_P, and I_P have a similar value while that of I_P is slightly higher than the values of V_P/ρ and V_P. However, it is obvious that in Figure 11 the value of I_P is higher than the values of V_P/ρ and V_P. According to Figures 10 and 11, ν and I_P are sensitive to both porosities and can be considered to build the rock-physics templates.

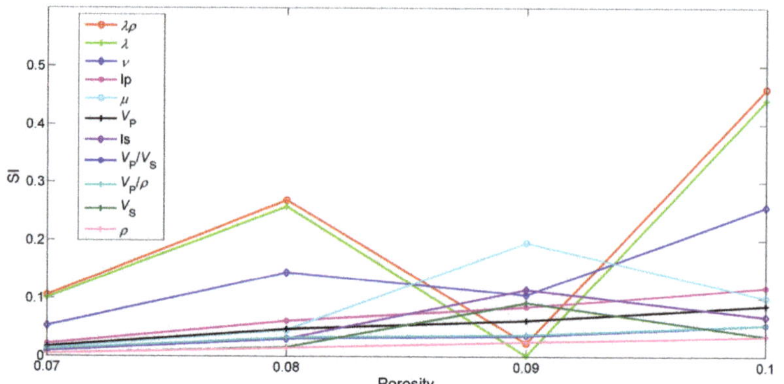

Figure 11. Sensitivity indices varying with the total porosity, based on data from Well P.

5. Rock-Physics Templates

5.1. Modeling

Figure 12 shows the RPTs with respect to total and microcrack porosities at ultrasonic and log frequencies, where the black and red curves isolines of total and microcrack porosities, respectively.

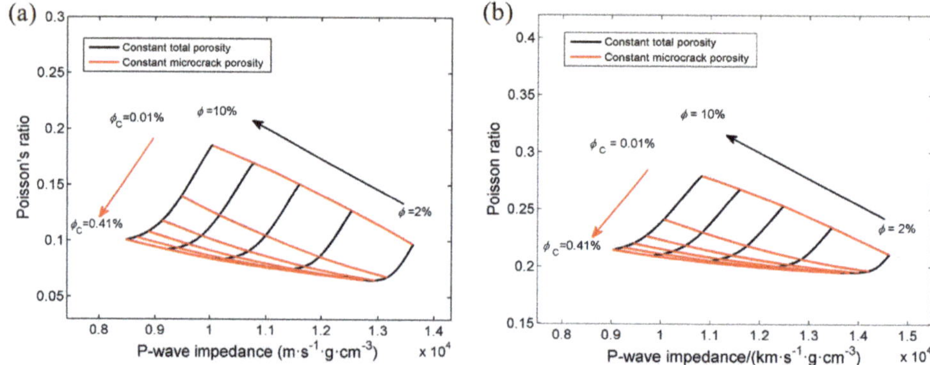

Figure 12. RPTs at ultrasonic (**a**) and sonic (**b**) frequencies.

5.2. Calibration

Ultrasonic and log data are used to calibrate the RPTs. Figure 13 displays an RPT at 1 MHz and the inclusion radius of 50 μm, where the color of the scatters indicates microcrack porosity, and the maximum value is 0.54%. The Poisson's ratio gradually increases with decreasing microcrack porosity; the plot shows that the template is in agreement with the ultrasonic data.

Figure 14 shows an RPT at 10 kHz and the inclusion radius of 200 μm, where the color bar indicates total porosity (the data are from the log data of well P). Most of the data have a porosity greater than 4%. With the increasing porosity, Poisson's ratio increases and P-wave impedance decreases, and the agreement between data and theory is good.

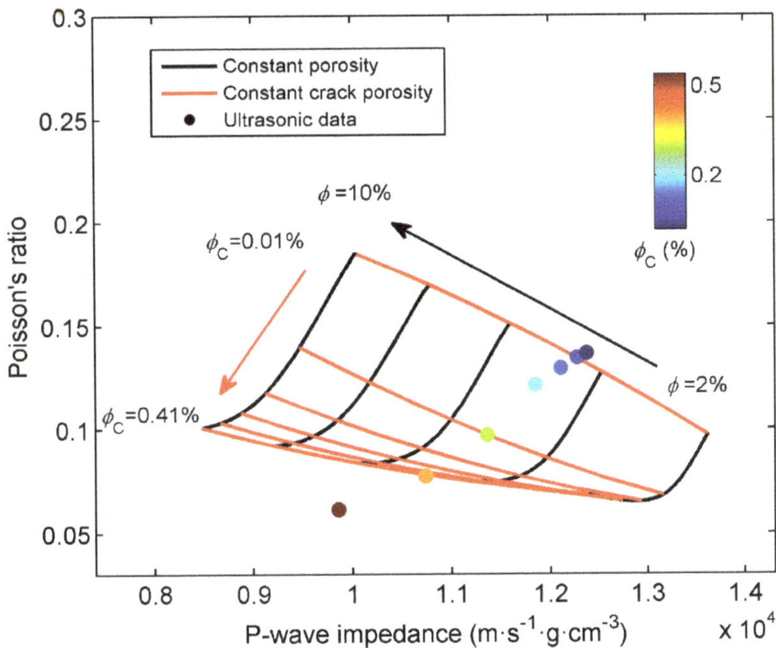

Figure 13. Ultrasonic data and corresponding RPT. The color bar indicates the microcrack porosity.

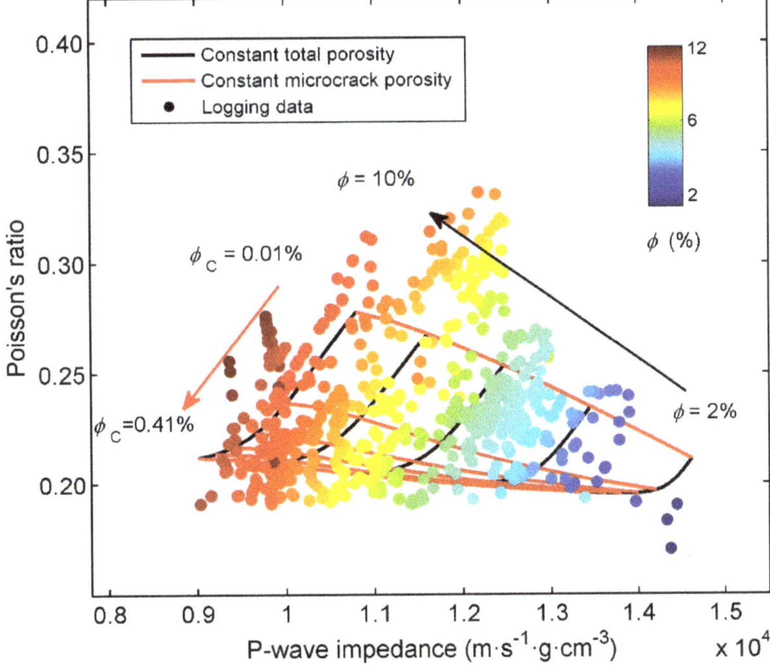

Figure 14. Log data and corresponding RPT. The color bar indicates the total porosity.

6. Microcrack Estimation

Based on approximated reflection coefficients [53], a pre-stack AVA inversion [54–56] is adopted to obtain the Poisson ratio and P-wave impedance and build the templates. Then, the microcrack porosity is estimated, with data points outside the template considered non-reservoir.

Figure 15 shows the P-wave impedance and Poisson's ratio obtained from a seismic survey line. Well M shows high P-wave impedances at the top and low P-wave impedances at the bottom of the target layer, and high Poisson's ratio in the middle. On the other hand, Well F shows the low impedance and Poisson's ratio. Figure 16 shows the total and microcrack porosities predicted for the survey. Both porosities are high at Well M and low at Well F. The gas production reports of wells M and F are 100.5×10^4 m^3 per day and 4.4×10^4 m^3 per day, respectively, indicating that the predictions are consistent with the reports.

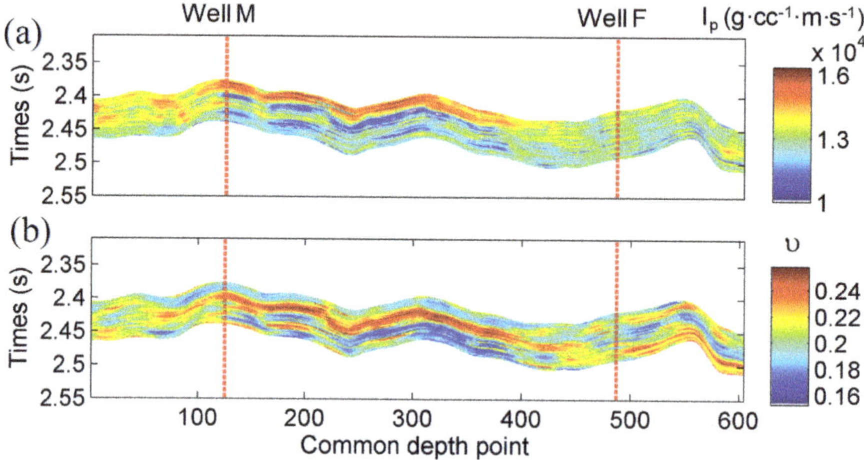

Figure 15. 2-D seismic profiles of P-wave impedance (**a**) and Poisson's ratio (**b**).

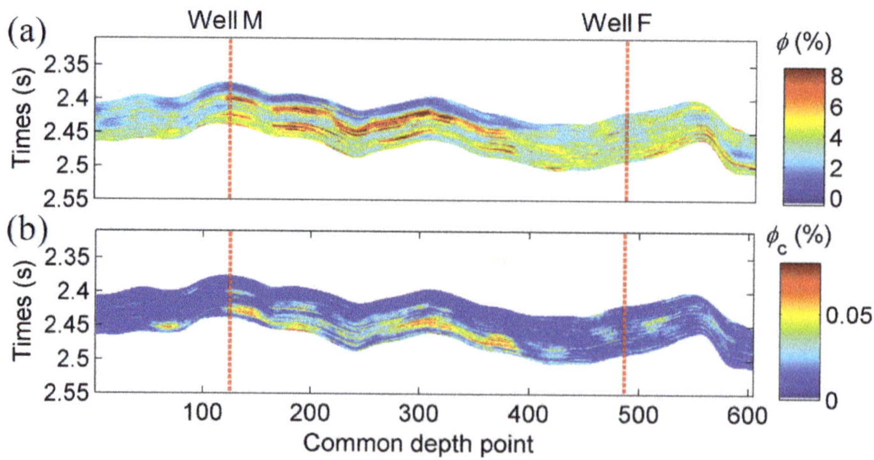

Figure 16. Inversion results of total (**a**) and microcrack (**b**) porosities corresponding to the data of Figure 15.

Figures 17 and 18 show results for another seismic survey line. Wells C and P show high P-wave impedances in the middle section, while Poisson's ratio is low throughout the target layer around Well C. This well exhibits low total and microcrack porosities (see Figure 18). Although Well P shows a higher total porosity, the microcrack porosity is low. Well J shows the best potential for gas production with high total and microcrack porosities. The reports show that Well C produced almost no gas with non-industrial gas production flow, while Wells P and J produce 2.14×10^4 and 6.4×10^4 m^3 gas per day, respectively. The predictions of the porosities are consistent with the production status.

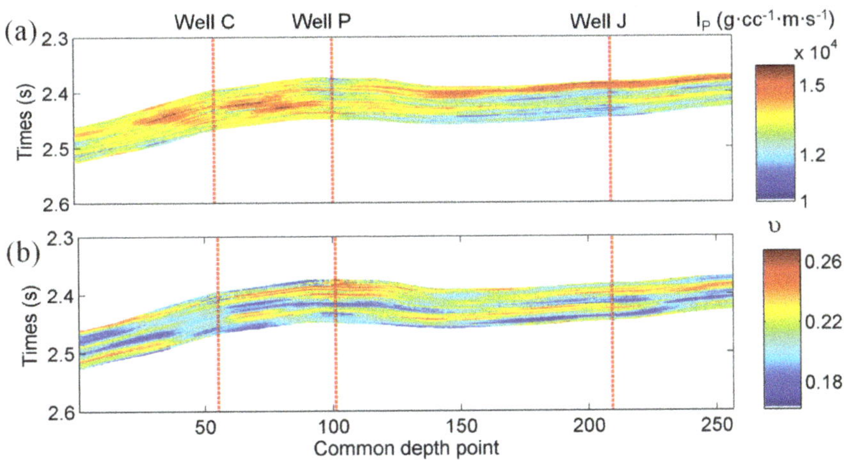

Figure 17. 2-D seismic profiles of P-wave impedance (**a**) and Poisson's ratio (**b**) corresponding to a second survey.

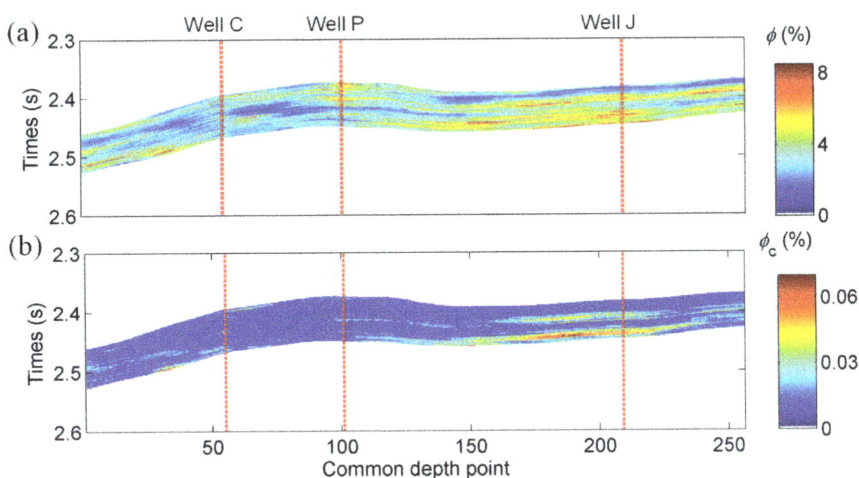

Figure 18. Inversion results of total (**a**) and microcrack (**b**) porosities, corresponding to the data of Figure 17.

3D seismic slices of the work area of ~471 km^2 are produced from the RPTs. According to the daily gas production rate, ten wells are classified into three categories, namely, extremely low (less than 1×10^4 m^3), low ($1 \times 10^4 - 7 \times 10^4$ m^3) and high (higher than 7×10^4 m^3) gas production wells. Figures 19 and 20 show the corresponding maps of total and microcrack porosities, respectively.

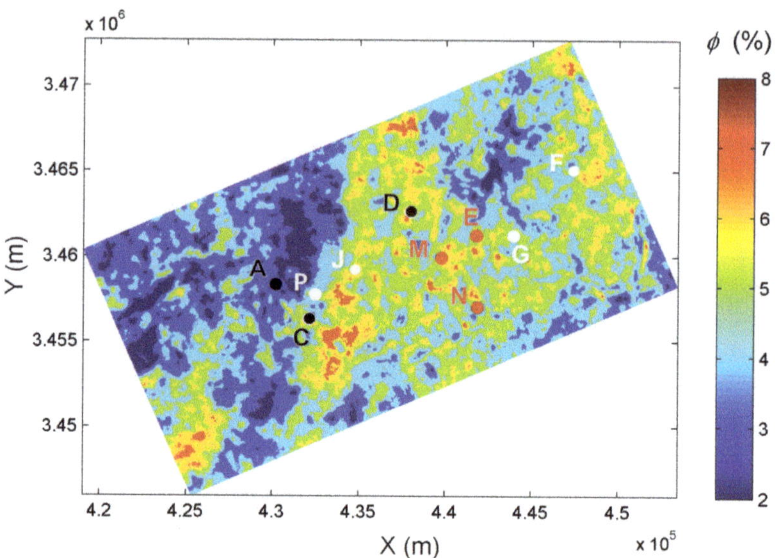

Figure 19. 3D slice of reservoir total porosity. Black, white, and red circles indicate extremely low, low, and high gas production wells, respectively.

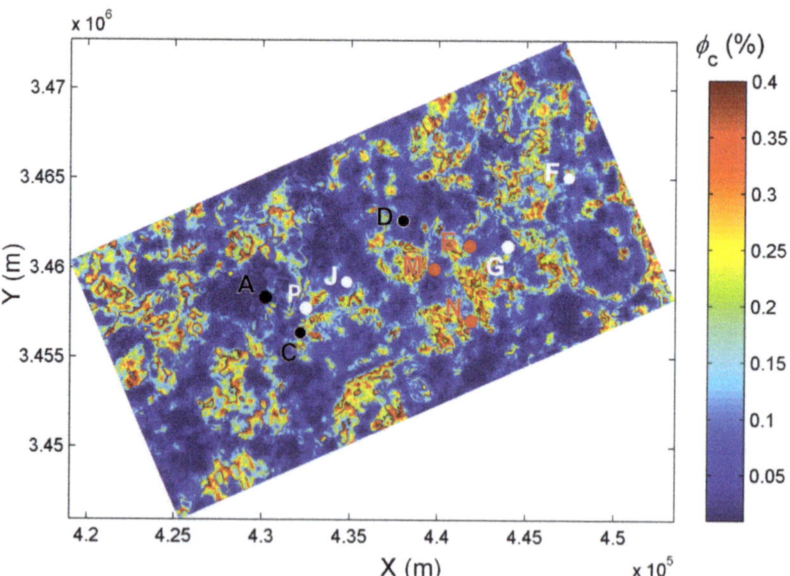

Figure 20. 3D slice of reservoir microcrack porosity. Black, white, and red circles indicate extremely low, low, and high gas production wells, respectively.

The predictions show that Well A is located in an area with extremely low total and microcrack porosities. The total porosity of Wells C and D are higher than that of Well A, but their microcrack porosities are low. Wells F, G, J, and P are located in an area with low total and microcrack porosities, whereas Wells E, N, and M show high total and microcrack porosities. According to the production test report, Well A is a water well with no gas production, and Wells C and D produce almost no gas. The three wells are classified as the lowest production. Wells P, F, G, and J produce 2.14×10^4, 4.4×10^4, 5.76×10^4, and 6.4×10^4 m^3 gas per day, respectively, classified as low production. Wells E, N and M produce 17.66×10^4, 25×10^4, and 100.5×10^4 m^3 gas per day, respectively, and are classified as high production. The predictions basically agree with the reports, so that the RPTs can discriminate between low and high gas production reservoirs.

7. Conclusions

This study uses rock-physics templates to estimate the total and microcrack porosities of tight sandstone gas reservoirs. The models are based on the poroelasticity theory to obtain the elastic attributes at different frequency bands, namely, how Poisson's ratio, velocity, and other relevant attributes are affected by those porosities. Ultrasonic experiments are performed under full water and gas saturations, showing that the wave velocities increase with effective pressure. When the rock is fully saturated with gas, the ratio between the P- and S-wave velocities increases with effective pressure, while for water, an opposite trend is observed.

A sensitivity analysis shows that P-wave impedance and Poisson's ratio are most sensitive to both porosities. The attributes are considered to build the templates which are calibrated with experimental and log data. Finally, the templates are applied to seismic data to estimate the porosities, indicating that the predictions are consistent with the gas production reports.

At present, the tight sandstone gas reservoirs of Sichuan Basin have contributed to the major proportion of the total gas production output of the petroleum industries in China. However, the tight sandstone reservoirs generally exhibit the characteristics of low porosity, low permeability, high heterogeneity, and deep burial. There are difficulties in the precise prediction and detailed description of high-quality reservoirs based on the traditional geophysical exploration methods. In this work, we propose a multiscale rock physics modeling method by incorporating the effects of microcrack porosity on rock physics attributes. A workflow is presented to establish the relationships between rock physical properties and seismic wave responses, based on which the prediction of microcrack porosity is achieved in the 3D seismic survey work area. The application of the method is effective as a case study.

The proposed method can be applied in rock physics modeling based on the elastic/seismic responses of reservoir rock. It may be affected by the problem of multiplicity of solutions in the parameter inversion and reservoir prediction. Further studies may incorporate the wave attenuation in rock physics modeling, so that more attributes will be available for validating and calibrating the templates, and the influence by the multiplicity of solutions can be alleviated. Furthermore, this paper is focused on the studies on tight gas sandstone reservoirs, and for the different types of complex reservoirs, the approach should be extended and adjusted to achieve effective applications. Specifically, for those tight/shale oil reservoirs, the viscoelasticity characteristics of fluid have to be analyzed and incorporated in rock physics modeling, so that some reasonable results can be obtained.

For future studies, the different sensitivity analysis methods should be included to analyze the interactions between different attributes with respect to rock basic properties. Furthermore, the multi-phases of different types of microcracks, mesopores and fractures need be considered to establish a more realistic rock model, with which the in situ reservoirs can be better described and simulated, in combination with the latest techniques of digital rock physics.

Author Contributions: Conceptualization, J.B. and R.H.; Data curation, T.C.; Formal analysis, C.R., J.B., J.M.C. and R.H.; Funding acquisition, J.B.; Investigation, J.B., J.M.C., T.C. and R.H.; Methodology, C.R., J.B. and J.M.C.; Project administration, J.B. and T.C.; Resources, T.C.; Supervision, J.B.; Validation, C.R. and J.M.C.; Writing—original draft, C.R. and J.B.; Writing—review & editing, J.B., J.M.C. and R.H. All authors have read and agreed to the published version of the manuscript.

Funding: This research was funded by National Natural Science Foundation of China, grant number 41974123, Jiangsu Province Outstanding Youth Fund Project, grant number BK20200021, and Strategic Priority Research Program of the Chinese Academy of Sciences, grant number XDA14010303.

Institutional Review Board Statement: Not applicable.

Informed Consent Statement: Not applicable.

Data Availability Statement: The data relevant with this study can be accessed by contacting the corresponding author.

Acknowledgments: The authors are grateful to the National Natural Science Foundation of China (Grant no. 41974123), the Jiangsu Province Outstanding Youth Fund Project (Grant no. BK20200021), and the Strategic Priority Research Program of the Chinese Academy of Sciences (Grant no. XDA14010303).

Conflicts of Interest: The authors declare no conflict of interest.

Abbreviations

AVO	Amplitude variation with offset
BR	Biot–Rayleigh
DEM	Differential equivalent medium
RPT	Rock physics template
FSI	Fluid sensitivity indicator

Appendix A

The Biot–Rayleigh theory

The BR dispersion equation [28] yields the complex wave number k,

$$
\begin{vmatrix}
a_{11}k^2 + b_{11} & a_{12}k^2 + b_{12} & a_{13}k^2 + b_{13} \\
a_{21}k^2 + b_{21} & a_{22}k^2 + b_{22} & a_{23}k^2 + b_{23} \\
a_{31}k^2 + b_{31} & a_{32}k^2 + b_{32} & a_{33}k^2 + b_{33}
\end{vmatrix} = 0,
\tag{A1}
$$

where

$$
a_{11} = A + 2N + i(Q_2\phi_1 - Q_1\phi_2)x_1, \; b_{11} = -\rho_{11}\omega^2 + i\omega(b_1 + b_2),
$$

$$
a_{12} = Q_1 + i(Q_2\phi_1 - Q_1\phi_2)x_2, \; b_{12} = -\rho_{12}\omega^2 - i\omega b_1,
$$

$$
a_{13} = Q_2 + i(Q_2\phi_1 - Q_1\phi_2)x_3, \; b_{13} = -\rho_{13}\omega^2 - i\omega b_2,
$$

$$
a_{21} = Q_1 - iR_1\phi_2 x_1, \; b_{21} = -\rho_{12}\omega^2 - i\omega b_1,
$$

$$
a_{22} = R_1 - iR_1\phi_2 x_2, \; b_{22} = -\rho_{22}\omega^2 + i\omega b_1,
\tag{A2}
$$

$$
a_{23} = -iR_1\phi_2 x_3, \; b_{23} = 0,
$$

$$
a_{31} = Q_2 + iR_2\phi_1 x_1, \; b_{31} = -\rho_{13}\omega^2 - i\omega b_2,
$$

$$
a_{32} = iR_2\phi_1 x_2, \; b_{32} = 0,
$$

$$
a_{33} = R_2 + iR_2\phi_1 x_3, \; b_{33} = -\rho_{33}\omega^2 + i\omega b_2,
$$

and

$$
x_1 = i(\phi_2 Q_1 - \phi_1 Q_2)/Z, \; x_2 = i\phi_2 R_1/Z, \; x_3 = -i\phi_1 R_2/Z,
$$

$$
Z = \frac{i\omega\eta\phi_1^2\phi_2\phi_{20}R_0^2}{3\kappa_{10}} - \frac{\rho_f\omega^2 R_0^2\phi_1^2\phi_2\phi_{20}}{3\phi_{10}} - \left(\phi_2^2 R_1 + \phi_1^2 R_2\right),
\tag{A3}
$$

where ω is the angular frequency; ϕ_{10} and ϕ_{20} are the local porosities of intergranular pores and microcracks, respectively; and ϕ_1 and ϕ_2 are the corresponding absolute porosities. ρ_f, η, and κ_{10} are the fluid density, the fluid viscosity, and rock permeability, respectively; A, N, Q_1, R_1, Q_2, and R_2 are elastic parameters; R_0 is the radius of inclusion; $\rho_{11}, \rho_{12}, \rho_{13},$